ACUA
Underwater Archaeology
Proceedings
2023

edited by

Sarah E. Holland and Bert Ho

Dave Ball, Series Editor

50 YEARS OF ACUA
1973-2023

AN ADVISORY COUNCIL ON UNDERWATER ARCHAEOLOGY PUBLICATION

The Editorial Team extends its many thanks to José Bettencourt and Marc-André Bernier for the Portuguese and French translations of all paper abstracts within this volume.

Made possible in part through the support of the Society for Historical Archaeology and the PAST Foundation.

SOCIETY *for*
HISTORICAL
ARCHAEOLOGY

Cover Image: Fishing boats off the Rock of Lisbon, 24 December 1856
Painted by Edward Gennys Fanshawe
Image reprint persmission generously provided by the
National Maritime Museum, Greenwich, London, England.
(© National Maritime Museum, Greenwich, London)

Foreword

Revisiting Global Archaeologies

Maritime archaeology is a specialized branch of archaeology and holds immense importance for societies by unlocking hidden treasures and untold stories either submerged beneath the world's oceans, seas, and waterways, or in close relation with those environments. This interdisciplinary field combines theories and methods fundamental for the study of archaeology, history, anthropology, and marine science to explore, protect, study, and share humanity's maritime heritage. Ships, ports, landscapes, and artifacts in close relation to maritime dynamics provide valuable insights into ancient civilizations, trade routes, navigation techniques, and cultural exchanges. By investigating shipwrecks, submerged cities, and coastal and underwater archaeological sites, maritime archaeologists work to piece together fragments of history, shedding light on different ontologies, significant events, cultures, identities, and technological advancements. This knowledge enriches our understanding of human history, fills gaps in the historical record, and enhances the accuracy and comprehensiveness of our narratives.

The ultimate objective of maritime archaeology is humanity and its relation to itself and to the water, in a movable, ever-changing context. We look upon the coast and underwater environments as unique repositories of cultural heritage. The social value of maritime archaeology lies in documenting, conserving, and managing archaeological sites, as well as interpreting, discussing, and sharing knowledge, ideas, and opinions about what we believe to have been the past. We want to explore the larger themes and construct narratives related to the intellectual environment of the time under study. The interaction between popular culture and super-structures, such as official religious doctrines, the dynamics of power and authority, the social structure of rural and urban communities, and the diversity of convictions and opinions in each population. The annual Society for Historical Archaeology (SHA) conference provides an ideal opportunity for professionals to come together and explore the rich potential of maritime archaeological sites, analyses, and investigations.

In our very biased opinion, we believe that Lisbon, Portugal, was the perfect choice for the 2023 annual meeting of the SHA. As the capital city of Portugal, it has been a gateway to the sea for a long time. Located at the mouth of the Tagus River, Lisbon played an important role in globalization beginning in the 15th century. A hub for navigation and maritime trade, Lisbon became a center of maritime exploration and shipbuilding and played a pivotal role in the Age of Navigation. Its enduring legacy stands as a testament to human curiosity, courage, and the quest for new horizons, but also all the negative aspects that came with that endeavor. During the 15th and 16th centuries, Portuguese explorers embarked with sailors from other European countries on ambitious voyages, seeking new trade routes, wealth, and knowledge of the world beyond Europe. The voyages of Bartolomeu Dias and Vasco da Gama opened new routes to Africa, India, and the rest of the world.

As much as the rest of the human adventure, the story of the European expansion is a story of violence, conquest, and subjugation of other peoples and cultures, but it is also a story of commerce, sophistication, scientific discoveries, exchanges of ideas, techniques, tastes, sounds, and artistic styles, of encounters, and the first ethnographies. One of the key factors that propelled Lisbon's navigation endeavors may have been the development of the caravel, a small, versatile, and seaworthy vessel. Nobody knows for sure how they were made, what they looked like, how they were rigged, or what the range of their tonnages were, but it seems that these vessels were small, cheap, and had a shallow draft yet were still maneuverable and fast. They played a major role in the 15th-century exploration of the Atlantic coast of Africa.

Advancements in cartography, navigation instruments, and maritime techniques, attracted adventurers from Europe, the Maghreb, and the Middle East, and empowered Lisbon's sailors to venture into uncharted territories, expanding the boundaries of human knowledge and the understanding of the world. In the 16th century, Lisbon became a melting pot of cultures, attracting merchants, explorers, and sailors from around the globe. The influx of wealth from maritime trade transformed Lisbon into a thriving cosmopolitan city, fostering a vibrant exchange of goods, ideas, and knowledge. The city's markets, such as the Ribeira Market, became vibrant centres of commercial activity, facilitating the trade of exotic spices, silks, precious metals, and other valuable commodities. Lisbon's historical connection to navigation and maritime exploration has left an indelible mark on the city's culture, architecture, and collective memory. Monuments and landmarks, such as the Tower of Belém, pay homage to Lisbon's maritime heritage and serve as reminders of the courage and ambition of those who set sail from its shores.

The city's maritime museums, such as the Maritime Museum of Lisbon, where SHA's opening conference reception occurred, provide invaluable insights into the navigational tools, ship models, and artifacts from this transformative period in history, reminding us that Lisbon's maritime heritage continues to be celebrated and cherished, serving as a testament to humanity's insatiable curiosity, resilience, and the enduring quest to unravel the mysteries of the seas. It was an honor and a privilege to receive the 2023 annual meeting of the SHA in Lisbon. In an increasingly interconnected world, scientific progress depends heavily on global cooperation and collaboration. International science meetings serve as vital platforms that bring together researchers, scientists, and experts from diverse backgrounds, embodying the exchange of knowledge, ideas, and experiences. These gatherings play a pivotal role in advancing scientific understanding, promoting interdisciplinary research, and tackling the complex challenges facing humanity. One of the primary benefits of international scientific meetings is their ability to facilitate collaboration and networking opportunities on a global scale. These events provide scientists with a unique platform to establish and nurture professional relationships with colleagues from different countries and institutions. Such connections often lead to fruitful partnerships and research collaborations that transcend geographical boundaries.

Scientific progress thrives on the dissemination of knowledge, and SHA has been more of a community of scholars where students get opportunities to present their research in a welcoming and nurturing environment. International science meetings serve as conduits for the exchange of research findings, cutting-edge technologies, and innovative methodologies, but SHA adds a non-competitive dimension, empowering students and subjecting them to the critique of their peers in a protective environment. If anything makes SHA different from other international meetings is that it is the least hierarchical and most relaxed and welcoming of them all.

Scholars and students present their work together, through oral presentations, poster sessions, and panel discussions, providing a platform to share their findings, methodologies, and insights. This open sharing of knowledge stimulates scientific discourse, sparks intellectual debates, and enables researchers to stay abreast of the latest developments in their respective fields. Moreover, exposure to a wide range of research topics and perspectives often sparks interdisciplinary collaborations and encourages the cross-pollination of ideas, leading to breakthroughs that may not have been possible otherwise. Conference evenings are famous for the discussions that occur and the projects they generate, the ideas they circulate, and the contacts they provide.

Many of the pressing challenges facing humanity, such as climate change, pandemics, and sustainable development, require global cooperation and interdisciplinary approaches. International science meetings provide a space where experts from diverse scientific disciplines can come together to address these complex problems. SHA should strive now to foster even more interdisciplinary dialogue and collaboration and make its annual meetings a feast of ideas, methodologies, and technologies from various fields. The resulting multidisciplinary approach will enhance even further the effectiveness of problem-solving and promote holistic solutions.

SHA also offers a platform for policymakers, industry leaders, and scientists to engage in discussions and explore strategies for the application of scientific knowledge to real-world challenges. As a melting pot of cultural diversity, SHA has tried to attract scholars and students from diverse backgrounds and continents and encourage them to interact and learn from one another. SHA gatherings provide an opportunity to appreciate and understand the rich tapestry of global scientific traditions, practices, and perspectives. Facilitating intercultural exchanges, the SHA meetings already foster an inclusive and collaborative scientific community that transcends national boundaries, but we should try to expand its reach and include a far larger pool of scholars.

The SHA Lisbon 2023 conference was a successful event, with a total of 956 participants. Precisely 741 abstracts were submitted (198 underwater and 543 terrestrial); these were divided into 64 symposiums (45 terrestrial and 14 underwater)—Lisbon introduced the concept of Open Symposia at the SHA—and 25 general sessions (15 terrestrial and 8 underwater). Seven discussion forums (2 terrestrial and 5 underwater) took place and there were also 10 roundtable luncheons, four tours, and four workshops.

This volume collects 20 papers resulting from an equal number of presentations in the underwater archaeology sessions. The geographical, chronological, and thematic diversity is outstanding. Several texts present micro-studies on ships and shipwrecks in Europe, Africa, South America, and North America, many of them using the systematic study of written sources and laboratory research, demonstrating the importance of interdisciplinary dialogue in historical archaeology. Other studies look at the landscape, an increasingly present subject in maritime archaeology, considering its material dimensions, submerged and on the coast, tangible and intangible, but also the development of training programs for its study. Others analyze specific material categories of the nautical world, demonstrating the importance that certain objects had in the formation of maritime identities. Also of note are studies that have as their background the management of maritime archaeological heritage, which is vulnerable to human activities and climate changes, and thus in need of urgent monitoring and conservation programs.

Overall, this volume summarizes how dynamic, challenging, and inclusive maritime archaeology has become; however, we would like to leave a word for the future. Maritime archaeologists live in a world of constant challenges. Oceans, rivers, and seas are some of the most important factors of mobility and resources and fundamental for a sustainable future. It is important that we add to our main concerns and research strategies, the many problems generated by accelerated processes and fight for a better future.

Acknowledgments

The conference was only possible due to the support of NOVA University of Lisbon. This institution permitted the Society for Historical Archaeology to use the space in their College of Humanities and Social Sciences (a different concept than the normal hotel venue), where the sessions took place. But also, the Navy Museum, who offered us their magnificent space for the off-site reception.

Felipe Castro, Centro de Arqueologia Marítima da Universidade de Coimbra, Figueira da Foz Campus
José Bettencourt, Centro de Humanidades, Faculdade de Ciências Sociais e Humanas da Universidade Nova de Lisboa
Tania Casimiro, Centro de Ecologia Funcional – História, Territórios Comunidades, Faculdade de Ciências Sociais e Humanas da Universidade Nova de Lisboa

In Memoriam:

Peter Hitchcock

(15 March 1970 to 12 September 2023)

Underwater archaeologist, explorer, and conservator Peter Hitchcock passed away on September 13, 2023, at the age of 53. Peter graduated from Texas A&M University (TAMU) in 1993 with a Bachelor of Arts degree in Anthropology and continued his studies at the university, earning an M.A. from the Nautical Archaeology Program (NAP) in 2002. As a student in 1993, Peter worked with TAMU on sites in Mount Independent and Lake Champlain. In 1997, Peter participated in the Texas Historical Commission's *La Salle* Shipwreck Project on excavation of the 17th-century French colonial bark, *La Belle*, part of Robert Cavalier Sieur de La Salle's expedition to the Gulf of Mexico. As part of the conservation effort, he worked at TAMU's Conservation Research Laboratory from 1997 to 2002 as a member of the team tasked with the conservation and reassembly of the hull. Peter assisted in the design and creation of the carbon-fiberglass composite fiber support framework used in its exhibit at the Bullock Texas State History Museum.

From 2002 to 2005, Peter worked on the excavation of the Red River Wreck, identified as the 1838 shipwreck of the steamboat, *Heroine*. Peter was the assistant director on this project, conducted by TAMU in partnership with the Oklahoma Historical Society. Peter was involved in two early deep-water archaeological investigations led by TAMU—the Mica and Mardi Gras Shipwreck Projects—both significant as early deep-water testing and data recovery projects in the Gulf of Mexico; he was the project manager on the Mardi Gras Project. Peter's experience in deep-water exploration led to his career at TDI-Brooks International, Inc., an offshore geophysical exploration company specializing in multi-disciplinary oceanographic projects including mineral exploration and offshore windfarms. Peter worked at TDI-Brooks in College Station, Texas, from 2010 until his retirement as a Senior Project Manager in 2022. His wife Molly and their two children survive him.

In Memoriam:

James R. Pruitt

(17 March 1986 to 02 February 2023)

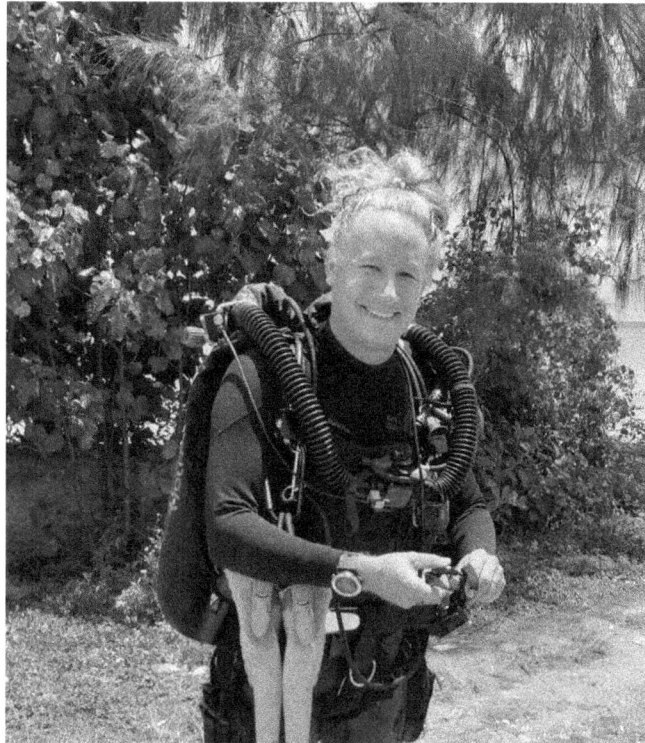

James R. "Jimmy" Pruitt, 36, was declared lost at sea on 02 February 2023, while scuba diving in the waters of Guam. Jimmy received his undergraduate degree from Middle Tennessee State University, majoring in Mass Communications. During his undergraduate studies, he moved to Japan as an exchange student and remained to teach English. He fell in love with Japan, spending much of his time traveling the country and SCUBA diving at every opportunity.

Jimmy returned to the United States to attend East Carolina University's (ECU) Maritime Studies Program (2013–2015). He was already a highly skilled diver and dive instructor, loved to discuss the technical side of diving, and delighted in any challenging underwater environment. During his two years at ECU, he took every opportunity to advance his career and received several awards and professional internships, including one with Naval History and Heritage Command.

Jimmy worked as a marine archaeologist employed by AECOM and the Commonwealth of the Northern Mariana Islands Historic Preservation Office. In April 2019, he temporarily joined Ships of Discovery on a Defense POW/MIA Accounting Agency project in Palau to recover the remains of service personnel lost in World War II. Jimmy's technical diving expertise and archaeological skills directly contributed to the recovery of two airmen, leading to their eventual identification and family reunification. In August 2020, Jimmy moved to Guam where he worked as an archaeologist for the Department of Defense, Naval Base Guam.

As his friends and colleagues can attest, Jimmy was full of life, quick with a joke and a smile, and took on every challenge with a lightheartedness that was both contagious and inspiring. His wife Sakura and their three children survive him. Jimmy had the mind of a scholar, the soul of an explorer, and the loving heart of a husband, father, son, brother, and friend. Good people leave big holes and Jimmy is, and will continue to be, much missed.

Looking to Shipwrecks: First Results of SUNK, a Research Project on Tagus Mouth Modern-Age Underwater Sites

José Bettencourt, Jorge Freire, Augusto Salgado, António Fialho

Research conducted on the mouth of the Tagus River since 2018 included monitoring the Bugio 1 (c. 1700) shipwreck and the survey of a large area. An older wreck (Bugio 2) was also recorded, which is probably the remains of the Portuguese nau São Francisco Xavier lost when returning from India in 1625. Furthermore, six other Modern-Age shipwrecks were discovered, the most outstanding being Parede 1 and Carcavelos 12, featuring artillery ordnance and ammunition, dating from the early 18th century or late 19th century. Project SUNK - Naufrágios modernos da Barra do Tejo (Early Modern Shipwrecks under Tagus Mouth) aims to contextualize these sites in terms of transoceanic navigation.

A investigação realizada na foz do Tejo desde 2018 incluiu a monitorização do naufrágio Bugio 1 (c. 1700) e o levantamento de uma grande área na periferia. Foi registado um naufrágio mais antigo (Bugio 2), que provavelmente corresponde aos restos da nau portuguesa São Francisco Xavier, perdida quando regressava da Índia em 1625. Foram ainda descobertos seis outros naufrágios da Idade Moderna, destacando-se o Parede 1 e o Carcavelos 12, com peças de artilharia e munições, datados de finais do século XVIII ou início do século XIX. O projeto SUNK – Naufrágios modernos da Barra do Tejo – pretende contextualizar estes sítios na navegação transoceânica.

Les recherches menées sur l'embouchure du Tage depuis 2018 comprenaient la surveillance de l'épave Bugio 1 (vers 1700) et le levé d'une vaste zone. Une épave plus ancienne (Bugio 2) a également été documentée, qui est probablement les restes du nau portugais São Francisco Xavier perdu à son retour d'Inde en 1625. En outre, six autres épaves de l'ère moderne ont été découvertes, les plus remarquables étant Parede 1 et Carcavelos 12, avec des pièces et des munitions d'artillerie, datant du début du 18ème siècle ou de la fin du 19ème siècle. Le projet SUNK - Naufrágios modernos da Barra do Tejo (Épaves modernes sous l'embouchure du Tage) vise à contextualiser ces sites en termes de navigation transocéanique.

Introduction

During the Modern Age (1500–1800), the Portuguese province Extremadura, particularly the port of Lisbon, became an important platform for intercontinental navigation, with ships sailing to and from ports all over the world, from the Mediterranean or northern Europe to Africa, America, and Asia (Costa 1997; Frutuoso et al. 2001). The Tagus estuary is an outstanding natural space known since antiquity. Yet, the approach to the port of Lisbon had some limitations due to the natural characteristics of the Tagus mouth—irregular bathymetry and morphology, combined with complex meteorological conditions. This increased navigational difficulties and limited the operation of the port, namely the departure and arrival times of the fleets sailing to, or returning from, different parts of the world (Boiça 1998). These difficulties, frequently mentioned in written sources, also explain the repeated occurrence of shipwrecks near the coast between the coastal site of Cabo da Roca and the island of Bugio.

Project SUNK (*Naufrágios Modernos da Barra do Tejo* [Early Modern Shipwrecks under Tagus Mouth]) aims to integrate the Modern-Age remains known in this area as elements of the maritime landscape of the Lisbon port complex that reflect ocean routes. As such, it is an extension of the research initiated in the scope of the *Projeto Carta Arqueológica Subaquática de Cascais* (PROCASC) (Cascais Underwater Archaeological Mapping Project) (Freire et al. 2020). This article presents the project, which started in 2021, and initial research results.

Research Methods and Principles

The SUNK project was designed to use a minimum impact methodology, allowing for a sustainable study of the Modern-Age underwater cultural heritage existing at the mouth of the Tagus River, including areas from Cabo Raso to the edges of the two access channels to the Tagus estuary. The study of these areas included systematic surveying based on historical and geophysical data analysis, whenever possible using information made available by other projects. To date, the multibeam surveys carried out by the Portuguese Environment Agency (APA) in the scope of the project *Programa de Monitorização da Faixa Costeira de Portugal Continental* (COSMO) (Coastal Monitoring Program of Continental Portugal) have been critical, covering the northern area of the river mouth, off the beaches of Parede and Carcavelos, and the surroundings of the Bugio lighthouse. The multibeam data provided the location of targets that revealed during diver inspection previously unknown archaeological sites, but which were also used in the site recording and monitoring.

All identified sites were systematically recorded using photogrammetry. As the sites are affected by sedimentary processes that erode and cover the remains over time, surveys are conducted on a yearly basis. During the first years, surveys were oriented and scaled based on targets placed before the photographic or video coverage, and georeferenced according to multibeam data. In 2022, the photogrammetric surveys were georeferenced with the underwater navigation, communications, and surveillance system (UWIS), which provides time-stamped location data for the photos used in 3D-modelling software. This non-intrusive strategy makes it possible to monitor the evolution of archaeological sites through direct observation and quantitative analysis of geophysical data and photogrammetry. It also provides data for the systematic characterization of the archaeological sites, as different distinct areas of the wrecks are exposed over time.

Initial Results

The study of available data and the first SUNK fieldwork in 2021 and 2022 contributed to the knowledge of the Bugio 1 and Bugio 2 shipwrecks (Figure 1). For Bugio 1, it was possible to map the two main deposits related to this wreck for the first time. In the northeast deposit are two iron anchors, one of which still has part of the wooden stock, and a concretion where fragments of faience plates (*pratos*) and olive jars (*botijas*) could be seen. In the southeast deposit, four exposed iron cannons, an anchor, and several deadeyes were observed among ballast stones, concretions, and cables, apparently made from piassava (*piaçava*) (*Attalea funifera or Leopoldinia piassaba*). These finds seem to confirm the dating proposed in a first assessment of the site (Monteiro et al. 2018). The presence of Portuguese faience decorated in cobalt blue, with concentric semi-circles on the plate rims, is particularly interesting and indicative of dating

to the second half of the 17th century. These materials appear to have close parallels in finds from the shipwrecks of *Santíssimo Sacramento* (1668) in Brazil (Neto 1977); from *Ponta do Leme Velho* in Cape Verde (1680–1700) (Gomes et al. 2014); or in the Portuguese frigate *Santo António de Tanna* (1697) excavated in Kenya (authors' observation).

The georeferencing of the Bugio 2 orthomosaics carried out in 2018 and 2019 provided an initial overall view and better understanding of the relationship with nearby Bugio 1. Bugio 2, probably the remains of the Portuguese nau *São Francisco Xavier* that sank there in 1625 when returning from India (Borges 2019), appears in four main areas (Figure 1).

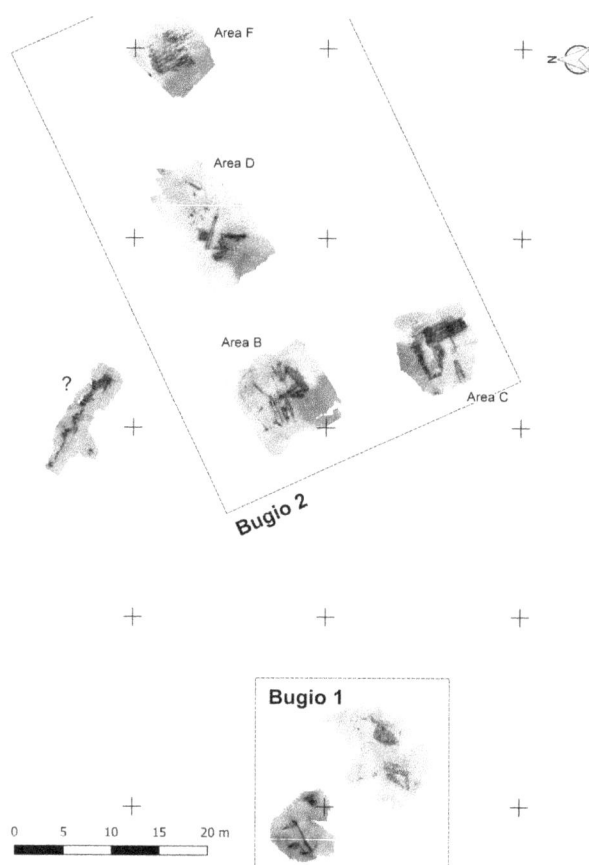

FIGURE 1. Orthomosaic of Bugio 1 and Bugio 2 - areas recorded between 2018 and 2021 (orientation to true north and georeferencing from 2018 multibeam data). (Figure by the primary author, 2023.)

In Area C, further south, a fragment of the central section of the hull is visible, with a keel section, central frames, and planking revealing the longitudinal axis of the hull and the shipwreck. Preliminary observation of this structure in 2018 indicated that it is a large ship, with close parallels to *Nossa Senhora dos Mártires*, another ship operating on the Portuguese Carreira da Índia (India Run) lost at the Tagus mouth in 1606 (Castro 2005).

In Area B, located 20 meters (m) north of the hull, five bronze cannons were exposed. Two artillery pieces date from before 1580, with the typical four rings instead of the usual dolphins, displaying the coat of arms of the Portuguese crown from the time of King Sebastião (1557–1578). The area also preserved scattered ship timbers, concretions, pepper, fragments of Chinese porcelain, and cowries.

A third zone (Area D), 20 m east of Area B, corresponds to another deposit with three bronze cannons (possibly of the colubrine or modified colubrine type), unidentified timbers, and numerous concretions, some featuring other artifacts. In this area, a stone mortar documented in situ in 2018 was removed by individuals not related to the project and without any archaeological documentation. The same happened with a concretion, whose precise location cannot be recovered. This piece was particularly interesting because it preserves a religious figure in ivory, fragments of pots from at least two types (Martaban and Tradescant), and Chinese porcelain (Figure 2). These are common trade goods on European ships involved in the Cape Route, the shipping route connecting the European coast to Asia's coast passing by the southern edge of Africa, opened by Portuguese sailors at the end of the 15th century.

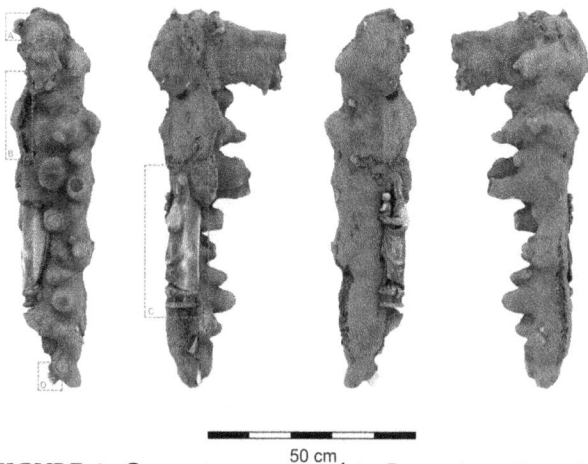

FIGURE 2. Concretions recovered in Bugio 2 in 2018: A: Tradescant pot handle; B: Martaban pot; C: ivory religious image; D: Chinese porcelain. (Figure by the primary author, 2022.)

Two more wooden structures, located further away, were also mapped. Area F, to the east, corresponds to another hull section, where porcelain and other materials can be seen between the frames. The last structure, of large dimensions, is located about 40 m north of Area B and seems to correspond to ship timber. So far, details of this structure have not been able to identify how it may relate Bugio 1 or Bugio 2.

Except for the hull, which is always visible, the other areas on Bugio 2 were not exposed after 2018, when the only available photogrammetric surveys were carried out. This could be indicative of sediment accumulation in the area, possibly favorable to the preservation of archaeological sites. However, an exploratory analysis of the multibeam data (comparing the 2018 and 2019 surveys) indicates that the area has a negative sediment balance, except for the area where archaeological remains were observed in 2018. In the case of Bugio 2, the archaeological remains seem to have acted as a sediment trap, influencing the local sedimentary processes at a scale that is impossible to assess at present.

In addition to the monitoring and documentation of known sites, such as Bugio 1 and Bugio 2, the work included the evaluation of targets identified in geophysical data. The observation of topographical anomalies detected in the multibeam surveys allowed the identification of six new shipwreck sites off the coast between São Julião da Barra and the town of Cascais. The following discussion highlights Parede 1 and Carcavelos 12, where the visible remains were more extensive. The others—Bugio 3, Bugio 4, Bugio 5 and Carcavelos 13—are at an early stage of investigation.

Parede 1 is shaped as a tumulus that stands out clearly from the surrounding area, at a depth of 11 m and reaching more than 1 m above the sandy bottom (Figure 3). The wreck is about 23 m long, with a north-south orientation, occupying a continuous area of 180 square meters (m²). The tumulus is composed of at least 16 iron cannons, the largest over 3 m in length, apparently stowed amidships. Deposits of concreted round iron shot, in clusters separated by spaces left by the bulkheads, divided the hold. At this stage, it is difficult to contextualize this wreck as no elements provide a clear date. The accumulation of cannons in the central section of the wreck, and the absence of others around the main site, suggest that this was cargo, which also included the round iron shot. The ship was, therefore, likely involved in merchant activities or logistical support to military operations. The size and formal characteristics of the iron cannon suggest a date in the 18th century or early 19th century.

Carcavelos 12 also corresponds to a military load (Figure 4); its top lies at a depth of 7 m. This wreck measures 11 m in maximum length, on a southeast-northwest direction, and is 7 m wide. The cargo is concreted and consists of wooden boxes, with round iron shot of at least two different calibers, and some iron shot as bulk cargo. Several alignments of unidentifiable pieces were also detected among the concreted mass. Six small-caliber bronze fire cannons were documented over the cargo, one of which was recovered and is currently undergoing conservation. A preliminary observation, before cleaning, showed that this object (CR12-21-001), measuring 0.91 m from base ring to muzzle, is a three-pound cannon. It was cast in 1808 or 1806 (the last number is difficult to read) and is a British gun, bearing the monogram of George III (1760–1820). The shipwreck must have occurred in the early 19th century.

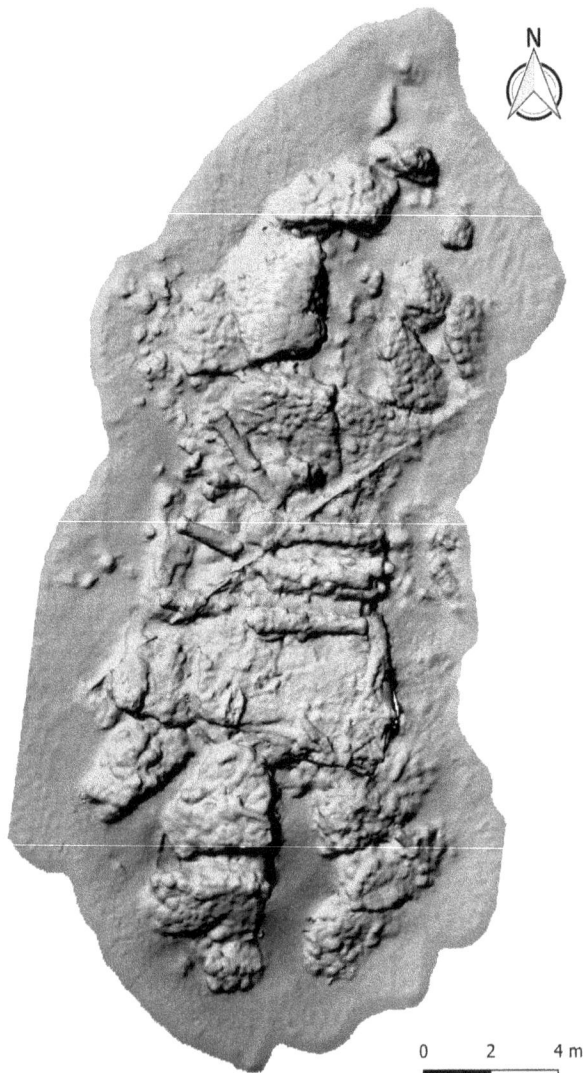

FIGURE 3. Digital surface model of Parede 1 as visible in 2021 (magnetic north). (Figure by the primary author, 2022.)

Final Remarks

The sites identified so far within the scope of the SUNK project document navigation between the early 17th and early 19th centuries and are markers of the commercial and military dynamics of the world in the Modern Age. The study of Bugio 2, probably the remains of *São Francisco Xavier*, which sank in 1625 when returning from India, is likely to contribute significantly to several research subjects. The first concerns shipbuilding. Structural remains of ships from the Portuguese India Run are rare. The research on *Nossa Senhora dos Mártires*, lost in 1606 in São Julião da Barra, is an outstanding case (Castro 2005). Bugio 2 shares characteristics with the latter, particularly the morphology and size of the keel or the planking, for example.

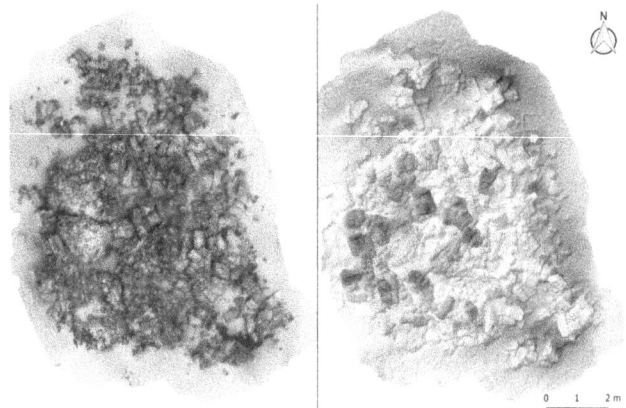

FIGURE 4. Orthomosaic and digital surface model of Carcavelos 12, as visible in 2022 (magnetic north). (Figure by the primary author, 2022.)

The second concerns equipment details, namely artillery, which represents between a quarter and a third of the total cost of a new ship ready to sail (Domingues 1998). Historical sources indicate the preferential use of bronze artillery, although they do not allow for a detailed analysis of their typologies, manufacture, or casting locations. Before the standardization of the foundry techniques and cannon types, on-board artillery was composed of a range of different calibers and typologies: large caliber pieces (*camelo*, *espera*, and *camelete*); half *espera*; stone falcons (*berços* and *falconetes*), and culverins (*colubrinas*) (Santos 1998). They could be manufactured in Portugal, in Portuguese India (*Estado da Índia*), reused from other vessels, or purchased abroad. Therefore, access to pieces with a well-defined archaeological context is a unique opportunity to analyze naval artillery. Once again, the Bugio 2 assemblage could make a decisive contribution to this research, especially when compared with those recovered in other Portuguese early modern contexts, namely *São Bento* (1554), *Santiago* (1585), *Santíssimo Sacramento* (1647), or *Nossa Senhora da Atalaia do Pinheiro* (1647) (Axelson 1985; Stuckenberg 1986; Santos 1986).

Thirdly, the archaeology of Bugio 2 could make an extraordinary contribution to the study of the cargos that circulated towards Europe. On the return trip, the origin, volume, and variety of goods transported on European ships were very diverse (Swahili coast and Mozambique, Hormuz, Iran, Arabia, India, Ceylon, Malacca, Siam, Pegu, Timor, China and Japan, and other places). The usual goods were spices, precious metals and stones, fabrics, wood, exotic products, and live animals (Russel-Wood 1992:177–199). Spices, cowries, porcelain, furniture, and ivory have already been documented on Bugio 2. The written documentation on *São Francisco Xavier* complements the information on cargo in transit, allowing for an integrated approach to written and archaeological sources, such as already applied in the study of *Nossa Senhora da Luz*, sunk in 1615 (Bettencourt 2005–2006). The material record to be retrieved from Bugio 2 could also be compared with data from *Nossa Senhora dos Mártires*, which revealed aspects on the ceramic cargo and life on board (Brigadier 2002; Coelho 2008).

The potential of Bugio 1 is more difficult to determine for now. However, this site's complementary inputs to the study of navigation dynamics in the early Modern Age are evident. As mentioned above, the considerably more recent Bugio 1 is dominated by the presence of artillery and several iron anchors. There are also some fragments of Portuguese faience and Iberian olive jars, along with piassava ropes, which were very popular in Brazil. This site could, therefore, be indicative of a link to the Atlantic, at a time which still lacks systematic studies on ships and shipwrecks. In Portugal, the only cases are the three, late-17th century Boa Vista ships, excavated in Lisbon (Bettencourt et al. 2018; Bettencourt et al. [2024]).

Parede 1 and Carcavelos 12, on the other hand, seem to correspond to merchant or support ships related to military operations in the 18th century or early 19th century, a period of significant political instability in the Iberian Peninsula. The characteristics and caliber of the Carcavelos 12 bronze cannons are consistent with land use, suggesting that this ship was heading to the port of Lisbon, with a military supply during the Peninsular War (*Guerra Peninsular*) (1807–1814), a conflict between the French Empire and the alliance of the United Kingdom of Great Britain and Ireland, the Spanish Empire, and the Kingdom of Portugal for control of the Iberian Peninsula during the Napoleonic Wars. The defense of Portuguese territory was possible due to the alliance with the United Kingdom, which sent military forces to support Portuguese troops in land defense, especially of Lisbon (Hall 2004). This hypothesis, almost certain for Carcavelos 12 and likely for Parede 1, opens new research perspectives in Portuguese underwater archaeology since the maritime dimension of the Peninsular War has been little investigated. It also allows a bridge with research on the same period conducted in Spain, where several shipwrecks have been studied in recent decades. For example, Deltebre I, a military transport ship of an English fleet that sank in the Ebro Delta, Catalonia, in the context of the French War (*Guerra del Francés*) (1808–1814) in the summer of 1813 (Vivar et al. 2016).

From a methodological point of view, several final aspects are noted. The exploration of geophysical data, obtained by other research programs with diverse investigation questions, revealed their potential for archaeological studies, with a minimum impact methodology. The combined use of multibeam data and photogrammetry has been particularly interesting, allowing the mapping of archaeological deposits quickly and accurately. These methods also support qualitative and quantitative monitoring of changes. The results obtained so far have revealed that the sites are exposed to complex site formation processes, which can affect their preservation and guide research methodologies and future management strategies. The SUNK project plans to broaden the use of these methodologies, seeking an interdisciplinary dialogue and a sustainable management of underwater cultural heritage.

Acknowledgments

The SUNK project is jointly developed by the Municipality of Cascais, CHAM–Centre for the Humanities - Universidade NOVA de Lisboa, the Portuguese Navy (namely through CINAV-EN and the Maritime Authority) and *Direcção Geral do Património Cultural* (DGPC). It is funded by the Municipality of Cascais and relies on the involvement of the technical and human resources of the other partners. This paper had the support of CHAM (NOVA FCSH / UAc), through the strategic project sponsored by FCT (UIDB/04666/2020). We would like to thank Armando Lucena for reviewing the English version.

References

AXELSON, ERIC
1985   Recent Identifications of Portuguese Wrecks in the South African Coast. *Estudos de História e Cartografia Antiga, Memórias*, Vol. 25. Instituto de Investigação Científica e Tropical, Lisboa: 43–61.

BETTENCOURT, JOSÉ
2005–2006 Os Vestígios da Nau *Nossa Senhora da Luz*: Resultados dos Trabalhos Arqueológicos (The Remains of the Ship Nossa *Senhora da Luz*: Results of Archaeological Works). *Arquipélago - História*, 2ª série, IX–X: 231–273.

BETTENCOURT, JOSÉ, INÊS COELHO, CRISTÓVÃO FONSECA, GONÇALO LOPES, PATRÍCIA CARVALHO, AND TIAGO SILVA
2018   Entrar e Sair de Lisboa na Época Moderna: Uma Perspectiva a Partir da Arqueologia Marítima (Entering and Leaving Lisbon in the Modern Age: A Perspective from Maritime Archaeology). In *Meios vias e Trajetos… Entrar e Sair de Lisboa. Fragmentos de Arqueologia de Lisboa 2 (Halfways and Routes… Entering and Leaving Lisbon. Fragments of Archeology of Lisbon 2)*, João Carlos Senna-Martinez, Ana Cristina Martins, Ana Caessa, António Marques, and Isabel Cameira, editors, Centro de Arqueologia de Lisboa e Sociedade de Geografia de Lisboa, Lisboa: 136–151.

BETTENCOURT, JOSÉ, PATRÍCIA CARVALHO, MÓNICA PONCE, TIAGO NUNES, GONÇALO LOPES, TIAGO SILVA, AND INÊS MENDES DA SILVA
[2024]   An Extraordinary Find? Boa Vista 5, A New Early Modern Ship Discovered in Lisbon Waterfront (Portugal). *Proceedings of the 16th International Symposium on Boat & Ship Archaeology*, 26 September to 01 October 2021, Zadar, Croatia.

BOIÇA, JOAQUIM MANUEL FERREIRA
1998   Zarpar e Arribar a Lisboa na Época da Navegação Moderna (Setting Sail and Arriving at Lisbon in the Age of Modern Navigation). In *Nossa Senhora dos Mártires". A Última Viagem (Nossa Senhora dos Mártires". The Last Voyage)*, Simoneta L. Afonso, Rafaela D'Intino, and M. Soromenho, editors, pp. 23–31. Expo'p8 – Pavilhão de Portugal e Editorial Verbo, Lisboa, Portucal.

BORGES, MARCO
2019   Viagem e Naufrágio de Uma Nau da Carreira da Índia: O Caso da *São Francisco Xavier* (1623–1625) (Voyage and Shipwreck of a Ship of the Indian Carrier: The Case of the *São Francisco Xavier*). *Fragmenta Historica – História, Paleografia e Diplomática 7 (Fragmenta Historica – History, Paleography and Diplomatics 7)*, Centro de Estudos Históricos da Universidade Nova de Lisboa, Lisboa: 71–91.

BRIGADIER, SARA R.
2002   *The Artifact Assemblage from the Pepper Wreck: An Early Seventeenth Century Portuguese East-Indiaman that Wrecked in the Tagus River.* Master's thesis, Texas A&M University, College Station, Texas. <https://oaktrust.library.tamu.edu/handle/1969.1/ETD-TAMU-2002-THESIS-B74 >. Accessed March 2023.

CASTRO, FILIPE
2005   *The Pepper Wreck: A Portuguese Indiaman at the Mouth of the Tagus River.* Texas A & M University Press, College Station, Texas.

COELHO, INÊS
2008   *A Cerâmica Oriental da Carreira da India no Contexto da Carga de uma Nau - A Presumível Nossa Senhora Dos Mártires (Oriental Ceramics from the Indian Era in the Context of a Ship's Cargo - The Presumed Nossa Senhora Dos Mártires).* Master's thesis, Faculdade Ciências Sociais e Humanas, Universidade Nova de Lisboa, Portugal.

COSTA, LEONOR F.
1997   *Naus e Galeões na Ribeira de Lisboa. A construção naval no sec. XVI para a Rota do Cabo (Indiamen and Galleons on the Lisbon Riverfront. Shipbuilding in the XVI century for the Cape Route).* Patrimómia Histórica, Cascais, Portugal.

DOMINGUES, FRANCISCO C.
1998   *A Carreira da Índia. The India Run.* Clube do Coleccionador dos Correios, Lisboa, Portugal.

FREIRE, JORGE, JOSÉ BETTENCOURT, AND AUGUSTO SALGADO
2020   Perdidos à Cista da Costa. Trabalhos Arqueológicos Subaquáticos na Barra do Tejo (Lost on the Coast. Underwater Archaeological Works at Barra do Tejo). *Arqueologia em Portugal 2020 - Estado da Questão.* Associação dos Arqueólogos Portugueses e CITCEM, Lisboa: 2059–2070.

FRUTUOSO, EDUARDO, PAULO GUINOTE, AND ANTÓNIO M. B.LOPES
2001   *O Movimento do Porto de Lisboa e o Comércio Luso-Brasileiro (1769–1836) (The Port of Lisbon Movement and the Luso-Brazilian Trade [1769–1836]).* Comissão Nacional para as Comemorações dos Descobrimentos Portugueses, Lisboa.

GOMES, MÁRIO V., TÂNIA CASIMIRO, AND JOANA GONÇALVES
2014   A Late 17th-Century Trade Cargo from Ponta do Leme Velho, Sal Island, Cape Verde. *International Journal of Nautical Archaeology* 44(1):160–172.

HALL, CHRISTOPHER D.
2004   *Wellington's Navy.* Chatham Publishing, London, England.

Monteiro, Alexandre, Jorge Freire, Flávio Biscaia, Paulo Costa, Marujo Gauthier-Bérubé, Pedro Patacas, and Sandro Pinto
2018   Notícia Preliminar da Descoberta de Dois Naufrágios na Entrada do Estuário do Tejo (Preliminary News of the Discovery of Two Shipwrecks in the Tagus Estuary Inlet). *Al-Madan Online*, IIª série, n.º 22, Tomo 1:166–170.

Neto, Ulisses de M.
1977   O Galeão *Sacramento* (1668): Um Naufrágio do Século XVII e os Resultados de Uma Pesquisa Arqueológica Submarina na Bahia (Brasil) (The Galleon *Sacramento* [1668]: A 17th Century Shipwreck and the Results of an Underwater Archaeological Survey in Bahia [Brazil]). *Navigator* 13:7–40.

Russel-Wood, Anthony J. R.
1992   *Um Mundo em Movimento: Os Portugueses na África, Ásia e América (1415–1808) (A World in Motion: The Portuguese in Africa, Asia, and America [1415–1808])*. Difel, Algés, Portugal.

Santos, Nuno V.
1986   *A Artilharia Naval e os Canhões do Galeão* Santiago *(The Naval Artillery and the* Santiago *Galleon Cannons)*. Academia da Marinha, Lisboa.

1998   Artilharia a bordo. In *Nossa Senhora dos Mártires". A Última Viagem (Nossa Senhora dos Mártires". The Last Voyage)*, Simoneta L. Afonso, Rafaela D'Intino, and M. Soromenho, editors, pp. 107–113. Expo'p8 – Pavilhão de Portugal e Editorial Verbo, Lisboa, Portucal.

Stuckenberg, Brian
1986   *Recent Studies of Historic Portuguese Shipwrecks in South Africa*. Academia de Marinha, Lisboa, Portugal.

Vivar Lombarte, G., Ruth Geli Mauri, and Thaís Torra Burgués
2016   El Vaixell Deltebre I: Resultats de les Excavacions Subaquatiques ` en un Vaixell de c`Arrega Militar (The Deltebre I Ship: Results of the Underwater Excavations in a Military Cargo Vessel. In: *200 Anys de la fi de la Guerra del Franc`es a les Terres de l'Ebre: Actes del Congr´es d'Historia i d'Arqueologia, Tortosa (200 Years of the end of the French War in the Lands of the Ebro: Proceedings of the Congress of History and Archeology, Tortosa)*, 16, 17 y 18 de maig de 2014, pp. 203–216. Onada Edicions.

. . . .

José Bettencourt
Centre for the Humanities and History Department (CHAM), FCSH
Universidade NOVA de Lisboa
Lisbon, Portugal

Jorge Freire
Câmara Municipal de Cascais, Portugal and
Centre for the Humanities (CHAM)
Universidade NOVA de Lisboa
Lisbon, Portugal

Augusto Salgado
Centro de História - Faculdade de Letras da Universidade de Lisboa and
Centro de Investigação Naval
Lisbon, Portugal

António Fialho
Câmara Municipal de Cascais and
Faculdade de Ciências Sociais e Humanas
Univeridade NOVA de Lisboa, Avenida de Berna, 26 C, 1069-061, Lisboa, Portugal

A Last Life: The Reuse of Ship Timbers on the Construction of River Waterfront on Rua D. Luís I (Lisbon, Portugal)

Mariana Mateus, José Bettencourt, Gonçalo Lopes, Nuno Neto, Raquel Santos, Luís Reis

The archaeological intervention between Dom Luís Street I and Boavista Street was an opportunity to document the west Lisbon riverfront, between the 17th and 20th centuries. Most of the relevant features recorded are related to the transformation of this space throughout the 19th century, including three wooden structures, revetments built to protect this area from the river. They were made by reusing ship timbers; one of these structures seems to correspond to the dismantling of a single ship. They can be related to what is known as clandestine embankments, made by private owners.

A intervenção arqueológica entre a Rua Dom Luís I e a Rua da Boavista constituiu uma oportunidade para documentar a frente ribeirinha ocidental de Lisboa, entre os séculos XVII e XX. A maior parte dos elementos registados estão relacionados com a transformação deste espaço ao longo do século XIX, incluindo três estruturas de madeira, cofragens construídas para proteger esta área do rio. Estas foram feitas através da reutilização de madeiras náuticas; uma destas estruturas parece corresponder ao desmantelamento de um único navio. As estruturas poderão estar relacionadas com aterros clandestinos, efetuados por proprietários privados.

L'intervention archéologique entre la rue Dom Luís I et la rue Boavista a été l'occasion de documenter le secteur riverain ouest de Lisbonne, entre les 17e et 20e siècles. La plupart des caractéristiques pertinentes enregistrées sont liées à la transformation de cet espace tout au long du 19e siècle, y compris trois structures en bois, des revêtements construits pour protéger cette zone de la rivière. Ils ont été fabriqués en réutilisant des bois de navire; l'une de ces structures semble correspondre au démantèlement d'un seul navire. Ils peuvent être liés à ce qu'on appelle des remblais clandestins, fabriqués par des propriétaires privés.

Introduction

The building project situated in the block between 18 Dom Luís I street and 51–59 Boavista Street, in Lisbon, included the construction of an underground car park, reaching a maximum depth of 7 meters (m) below street level (approximately 4 m below mean sea level [BMSL]). Archaeological work conducted between 2019 and 2020 in the southernmost part of the 460 square meter (m²) block resulted in the identification of several archaeological structures and deposits related to the riverfront (Mateus et al. 2021).

The maritime structures include a few wooden revetments used in the construction of Boa Vista landfill. Around 600 nautical timbers were reused, datable to the second half of the 18th century and the first half the 19th century. The archaeological deposits also yielded organic materials, such as shoe soles; ropes; different types of seeds; ceramics, such as Dutch and English kaolin pipes and an almost complete faience jug; some coins, and net weights of different types (Mateus et al. 2021).

The site was excavated by Neoépica, an archaeology company. The timbers were recorded in partnership with Centre for the Humanities (CHAM) -Faculty of Social Sciences and Humanities, NOVA University of Lisbon (NOVA FCSH). During fieldwork the timbers were assigned an individual reference; in a second phase, after being disassembled or removed from the site, they were cleaned and documented. Every single nautical timber was fully recorded, according to a set of parameters: identification, typology, dimensions, condition, and description of features and other relevant details (Bettencourt and Lopes 2021:8–9). These elements were analyzed by direct observation, recorded in individual sheets, including sketches. The archaeological documentation sought to obtain enough data to support a reconstitution of the shape of each timber to scale, also including technical aspects related to the conversion methods, the carpentry marks, the type and pattern of the fastening system, the form and typology of the scarfs, and the positioning and organization in the structure. Thus, the timbers were recorded by Structure from Motion (SfM) photogrammetry for all their faces. The more relevant and informative timbers were also recorded by means of a laser scan, to enable three-dimensional (3D) modelling. All the timbers were subsequently vectorized with CAD software (Bettencourt and Lopes 2021:9–12). The individual recording of these timbers allowed researchers to understand the typology of the elements reused in the structures. Samples were also taken for dendrochronology studies, following a partnership with the University of Coimbra. This article presents the contexts and a first approach to the study of the ship timbers.

The Ship Timbers and their Context

The oldest deposits documented concern the geological substrate, which corresponds to Miocene clays and limestones at approximately -2 m BMSL. The site was gradually occupied from the 16th century onwards. The first occupation phase (Phase 1) yielded evidence related to maritime activities on the beach. Later, several revetments were filled with reclamation dumps built to gain land from the Tagus River. A second phase of occupation (Phase 2) documents the conversion of this space into one or more production units or factories, possibly related to metallurgy and metal casting, from the mid-19th to mid-20th centuries (Mateus et al. 2021:28–31).

Phase 1 - First Occupation

The archaeological record pertaining to the first occupation phase featured loose remains and scattered nautical timbers related to the maritime landscape. They were found on the beach, over the levels of river muds, characterized by large amounts of malacological fauna, such as cockles (*Cerastoderma edule*) and oysters (*Ostreidae*). Material culture was very scarce, but some imported pottery has been identified, namely fragments of plain tin-glazed pottery from Seville, dating from the 16th to 17th centuries.

Some nautical timbers have also been identified, located primarily in the south and west quadrants, with north-south orientation (Mateus et al 2021:31–34). A keelson of large dimensions (RB59-792) stands out. It measured over 12 m in length, rectangular in cross-section, had a plain scarf at the preserved end and a series of iron nails regularly spaced along its entire length.

Phase 1 - Second Occupation

The second occupation phase featured a number of posts, without any organization or spatial pattern. These vertical elements had no wedge and were often supported by another piece of wood, usually chopped hull planks, which served as a base to provide greater stability (Mateus et al. 2021:35–38). Their construction was made by reusing 41 nautical timbers, mainly futtocks and floor timbers.

These deposits also yielded the scattered remains of a boat, Boa Vista 3 (BV3), located on the east and south corners between -1.07 and -1.77 m BMSL, disturbed by the construction of a wall. The timbers were interweaved with layers of silt. The stern was connected, sloping over the port side, with a north-south orientation. Some wooden blocks (two pulleys, RB59-500 and RB59-1015) and remains of rope rigging, mainly made of hemp, were also found (Bettencourt and Lopes 2021:13–21).

Phase 1 - Third Occupation

The third occupation corresponds to the construction of three wooden structures (611, 614, 615) (Figure 1). The first and oldest was a confined area (611) located in the west-central section of the site; the second, located further south and east (614) was north-south oriented; and the third deposit (615) was located in the southwest corner of the excavation area.

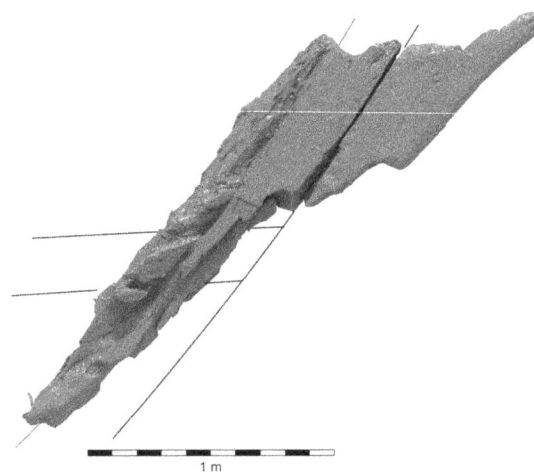

FIGURE 1. Phase 1 main features orthomosaic. (Figure originally published in Mateus et al. 2021, coauthors of the current publication.)

Structure (611) is a revetment made of posts and planks. The planks were laid out horizontally, overlapping each other, and were attached to the posts with iron nails. The planks consisted of reused planking or ceiling planks, from several ships of different sizes. In addition to the hull planks, three floor timbers, a bilge pump tube, an inner stern knee, and a heel were also reused in this structure (Mateus et al. 2021:43–75).

The three floor timbers (RB59-77 to RB59-79), together with four more recovered during the excavation, are an interesting set (Bettencourt and Lopes 2021:26–29) that could be part of the same ship (Figure 2, Group 1). All these floor timbers have similar characteristics, like semicircular limber holes; iron-only fastenings on the connection with the keel, futtocks, and the hull planks. Another set consisted of two floor timbers (RB59-190 and RB59-193, Group 2), with different characteristics, namely rectangular limber holes, and wooden treenails on the connection with futtocks (Bettencourt and Lopes 2021:26–29).

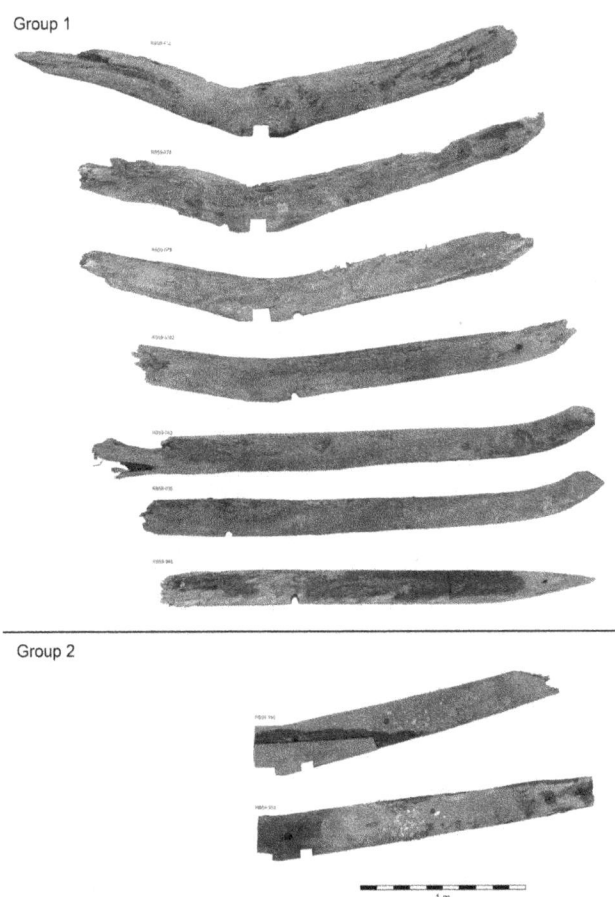

FIGURE 2. Floor timbers from Group 1 and Group 2. (Figure originally published in Bettencourt and Lopes, 2021, coauthors of the current publication.)

The second structure (614), north-south oriented, also consisted of post-like elements and planks, attached using mostly iron nails. The horizontal members were all hull planks (54 timbers), more than 5.50 m long, around 30.00 centimeters (cm) wide and 5.00 to 6.00 cm thick.

They show similar dimensions and fastening patterns, using iron nails and wooden treenails. Most of the wooden treenails have an incised X to expand the wood, which gave the ship's structure more strength and rigidity. Most of these elements had traces of caulking and abundant small copper nails placed at a short distance. These features are indicative of large wooden ships and suggest the use of copper sheathing, indicating a chronology after the last quarter of the 18th century. The use of copper sheathing became more common in the 19th century; several parallels are known around the world, such as the Mica shipwreck (Atauz et al. 2006:36–37), Baía da Horta 6 (Bettencourt et al. 2017b: 1999–2000), *Flower of Ugie* (Whitewright and Satchell 2011:551), *Rapid* (Staniforth 1985:32) and Slufter 4 (Adams *et al.* 1990:102–104).

The post-like elements corresponded mostly to futtocks and floor timbers (68 timbers). Two bilge pumps were also identified (RB59-159 and RB59-234). Bilge pump tube RB59-159 (Figure 3) is a 2.34 m long and 19.0 to 27.5 cm wide timber, with an octagonal section. Across its total length, the tube features a bore with 8.2 to 8.6 cm in diameter, where the valves worked and the water was pumped out. On one side is a notch for better accommodation on the pump well. A small groove is on one side and along the total length, 2.9 to 3.3 cm wide and 1.0 to 1.2 cm deep, to accommodate the chain. The lower end was chopped to facilitate water suction. The other pump tube (RB59-234) is similar. Both are also like the assemblage discovered at Boqueirão do Duro (Lopes et al. 2021:357–358), albeit smaller. Parallels can also be found in several pumps from 16th to 18th centuries shipwrecks, such as *Lomellina* (Guerout et al. 1989:49–61), *San Juan* (1565) (Loewen 2007:III-164–166), *Mary Rose* (McElvogue 2009:288–290), *Gagliana Grossa* (Batur and Rossi 2021:341–343), Studland Bay (Thomsen 2000:76–77), *Mortella III* (Cazenave 2020:100–108), *San José y Las Animas* (1733), *Machault* (1760) and *Defence* (1779) (Waddell 1985:246–247; Oertling 1996:24–29). They also appeared reused in port structures, for instance in London (Heard and Goodburn 2003:38–41).

The third structural feature (615), in the southernmost area, consisted of a group of wooden posts forming two alignments in east-west direction. The connection between these elements consisted of notches in the beams, while the posts were carved to accommodate this attachment. In some cases, the connections were additionally reinforced with iron nails. These elements should relate to the continuation of the structure to the south, namely with other timbers outside the excavation area (Mateus et al. 2021:43–75). All these structures were covered by multiple dumps, north-south and east-west oriented.

FIGURE 3. Bilge pump tube RB59-159. (Figure originally published in Bettencourt and Lopes, 2021, coauthors of the current publication.)

The larger posts consisted of reused nautical timbers, specifically a keel and a floor timber. The keel RB59-049 is 3.42 m long, 17.00 to 18.00 cm sided, and 26.00 to 31.00 cm molded. It features a triangular rabbet across both sides, measuring 3.5 cm in depth and 5.0 cm in height, where both garboards fitted, being reinforced by 0.5 cm square iron nails. On the top, the keel features square iron nails, 1.5 to 1.9 cm, used in the connection from floor timber to keel: they are arranged at regular intervals of 43.0 to 45.0 cm, which corresponds to the room and space between the floor timbers of the ship. In broad terms, this keel's dimensions are close to those recorded on the Boa Vista 1 ship (Bettencourt et al. 2021:23), a small/medium sized vessel, found nearby, about 70 m to the southeast (Lopes 2022:116–118).

The end section of the stem post of a wooden vessel (RB59-458, RB59-459 and RB59-460) (Figure 4) was found near another structural piece (615). The connection between these three elements consisted of a scarf more than 20 cm long, 15 cm wide, and 17 cm deep, reinforced by round iron bolts, 2 to 3 cm in diameter. The bow timbers display a triangular rabbet on each side, 10 cm high and 6 cm deep, to fit the hull planks; remains of at least three planks were present, covered by concretions originated by the iron nails used to attach them. Two of the timbers formed the bobstay piece of the head, fixed by iron fastenings. The dimensions of the bow, namely the stem post section, are around 17 to 23 cm molded and more than 34 cm high, indicating it was part of a medium-sized wooden vessel.

In addition, the same deposits preserved rigging pieces and nautical equipment: a block (RB59-369), a deadeye (RB59-502), four wooden sheaves and two fragments of oars (RB59-s/id7 and RB59-1003). The block and the deadeye show very common features from the mid-17th century onwards. The oar fragments, despite their poor condition, are very uncommon in archaeological contexts, especially on Lisbon's waterfront (Bettencourt and Lopes 2021:32–33).

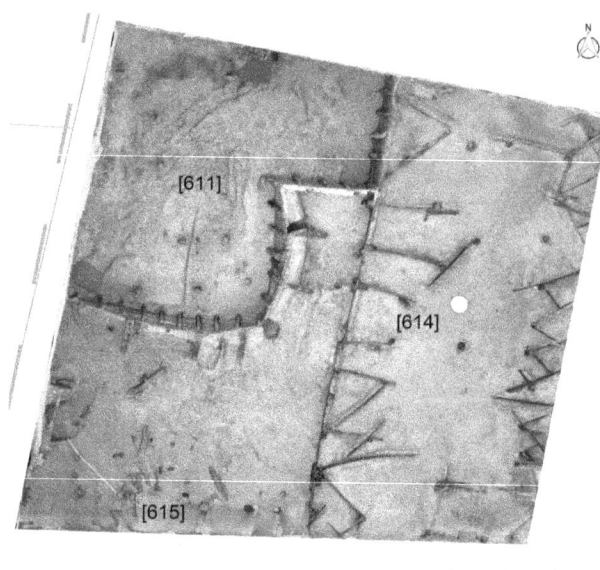

FIGURE 4. Stem post-timbers RB59-458, RB59-459, and RB59-460. (Figure originally published in Bettencourt and Lopes 2021, coauthors of the current publication.)

Phase 2

In the last occupation phase, the area was levelled to accommodate some walls, located in the central excavation area. A middle wall, with a north-south orientation, was erected along the eastern border of the wooden structure described above. This central wall was cut at its southernmost limit; its foundation extended beyond the excavation area. This structure must predate the opening of D. Luís I Street in the mid-20th century and would have changed after, before its extension to the south, eventually as far as present-day Av. 24 de Julho. Two parallel structures with an east-west orientation and another with a north-south orientation were found on the west side of this central wall, forming an enclosed area. Also related to this phase are some circular levels, namely cement sidewalks, demarcated areas of brick and mortar, and heavy machine bases, framed after the opening of D. Luís I Street (Mateus et al. 2021:76–104).

Conclusions

The archaeological intervention between Dom Luís I Street and Boavista Street revealed two different phases of occupation: the first is related to maritime uses and the second was industrial and can be associated with the "small private metal workshops" that existed "alongside the large state foundries" of the 19th century (Barreto 1981:482).

The first phase includes three wooden structures, two of them undoubtedly used as revetments to support reclamation dumps, probably illegal landfills predating the great public works of the Boavista landfill that began in 1855 (Araújo 1993: XIII-86). The north-south oriented structure (614) may also correspond to the foundations of a port structure, possibly a jetty or mooring facility perpendicular to the river. The available cartography supports this possibility and confirms the presence of a port-like structure at the site in the early 19th century. Identical wooden structures have been recorded worldwide, and they are very common in the city of London, which underwent several remodels along the coast, mainly between the 12th and 17th centuries (Schofield et al. 2018).

The first phase contexts yielded about 754 wooden pieces, 613 of which correspond to nautical timbers, mostly reused in the construction of the structures. The assemblage is heterogeneous and originates from several functional groups—longitudinal ship structure, frames, planking, rigging, and equipment—but research revealed an extensive reuse of planks and frames. They belong to multiple ships and boats, of different sizes and from different shipbuilding traditions. This pattern is similar to the Boqueirão do Duro site, located on the same street just 130-m west from the present site (Mateus 2018; Bettencourt et al. 2017a).

The nautical timbers suggest several strategies of reuse. The western boundary was made with two levels of hull planks still connected when they were removed, indicating that this is a section of a single ship. Some of the frames used as posts, mainly futtocks and floor timbers, were wedged to be driven into the foreshore, also show a clear coherence and similarity, leading the authors to believe that this structure was also made from a single ship's timbers.

The recycled nautical timbers are also an opportunity for detailed and extensive studies on shipbuilding and ships' equipment. For example, the current context has preserved items rarely found in archaeological contexts, such as the two bilge pump pipes, the rigging elements, and the oar fragments.

In addition, the systematic documentation of reusing and recycling strategies on waterfront contexts supports researching another stage of the life cycle of ships (Koivikko 2017:147–155). That research gives voice to the ships that, having lost their identity, ended their lives, but simultaneously became the means for the creation of other structures and objects (Leino 2013:136–138). The study of reused ship timbers also contributes to a better understanding of the appropriation of ship timbers and, subsequently, of the development of waterfront areas over the centuries. Lastly, it provides an approach to the concept of "blue cultural studies" (Mentz 2009:96–99). This concept emerged as a challenge to the hegemony of "green" in the universe of sustainability and ecology and seeks a more interdisciplinary and water-centered approach. Recently, it has resulted in another concept, more comprehensive and popular among the political-scientific community - the "blue humanities" (Brayton 2012; Mentz 2015). Indeed, by studying the reuse and recycling of nautical parts, it is also possible to characterize the sustainability of maritime communities, as well as their ecological concerns.

Acknowledgments

This paper had the support of CHAM (NOVA FCSH/UAc), through the strategic project sponsored by FCT (UIDB/04666/2020). We would like to thank Armando Lucena for reviewing the English version.

References

ADAMS, J., A. F. L. VAN HOLK, AND T.J. MAARLEVELD
1990   *Dredgers and Archaeology. Shipfinds from the Slufter.*
       Ministerie Van Welzijn. Volksgezondheid en
       Cultuur. Archeologie Onder Water, Rotterdam.

ARAÚJO, NORBERTO
1993   *Peregrinações em Lisboa. (Peregrinations in Lisbon.)*
       XIII Vol. 2ª edição. Veja, Lisboa.

ATAUZ, A. D., W. BRYANT, T. JONES, AND B. PHANEUF
2006   *Mica Shipwreck Project. Deepwater Archaeological
       Investigation of a 19th Century Shipwreck in the Gulf
       of Mexico.* Department of the Interior, Minerals
       Management Service, Gulf of Mexico OCS Region,
       New Orleans, LA.

BARRETO, JOSÉ
1981   Uma greve fabril em 1849. *Análise Social* (A Factory
       Strike in 1849. *Social Analysis*) XVII(67–68):
       479–503.

BATUR, KATARINA, AND IRENA RADIC ROSSI
2021   The 16th-Century Pump from *Gagliana Grossa*:
       Preliminary Results of Recording and Analysis.
       In *Proceedings of the 15th International Symposium
       on Boat and Ship Archaeology: Open Sea, Closed
       Sea - Local Traditions and Inter-regional Traditions
       in Shipbuilding*, Marseilles, Julia Boetto, Patrice
       Pomey, Pierre Poveda, editors, pp. 341–343.

BETTENCOURT, JOSÉ, AND GONÇALO C. LOPES
2021   *Rua da Boavista 59 – Relatório da análise de
       madeiras náuticas. (Boavista Street 59 - Report of the
       Nautical Wood Analysis.)* Centre for the Humanities,
       Centro de Humanidades e Neoépica, Lda. Lisboa.

BETTENCOURT, JOSÉ, GONÇALO C. LOPES, CRISTÓVÃO
FONSECA, INÊS PINTO COELHO, TIAGO SILVA, AND PATRÍCIA
CARVALHO
2017a A dimensão marítima do Boqueirão do Duro
       *(Lisboa): relatório preliminar do registo e avaliação
       das peças náuticas identificadas durante a construção
       dos novos escritórios da Fidelidade Europe Property,
       S.A (The Maritime Dimension of Boqueirão do Duro
       (Lisbon): Preliminary Report of the Registration and
       Evaluation of the Nautical Pieces Identified during the
       Construction of the New Fidelidade Europe Property
       Offices, S.A)* Centre for the Humanities, Centro de
       Humanidades, Lisboa.

BETTENCOURT, JOSÉ, TERESA QUILHÓ, CRISTÓVÃO FONSECA,
AND TIAGO SILVA
2017b Baía da horta 6 (BH-006): um provável naufrágio
       americano do século XIX (Horta Bay 6 (BH-006):
       A Probable 19th-Century American Shipwreck).
       Arqueologia em Portugal, Estado da Questão.
       Associação dos Arqueólogos Portugueses, Lisboa:
       1993–2009.

BRAYTON, DAN
2012   *Shakespeare's Ocean. An Ecocritical Exploration.*
       University of Virginia Press, Charlottesville, VA.

CAZENAVE DE LA ROCHE, ARNAUD
2020   *The Mortella III Wreck: a Spotlight on Mediterranean
       Shipbuilding of the 16th Century.* British
       Archaeological Reports, 2976, Oxford, UK.

GUEROUT, MAX, ERIC RIETH, AND JEAN -MARIE GASSEND
1989   Le navire Genois de Villefranche-un naufrage
       de 1516? (The Ship Genois de Villefranche - A
       Shipwreck in 1516?) *Archaonautica 9,* CNRS, Paris.

HEARD, K., AND D. GOODBURN
2003   Investigating the Maritime History of Rotherhithe:
       Excavations at Pacific Wharf, 165 Rotherhithe
       Street, Southwark, *MoLAS Archaeology Studies
       Series, 11*, London, UK.

KOIVIKKO, MINNA
2017   *Recycling Ships: Maritime Archaeology of the
       UNESCO World Heritage Site, Suomenlinna.*
       Finnish Maritime Archaeological Society, Vol. 1,
       Helsinki.

LEINO, MINNA
2013   Recycling Shipwrecks - Examples from the
       18th-century Fortress Island of Suomenlinna. In
       *Interpreting shipwrecks. Maritime Archaeological
       Approaches.* Jonathan Adams and Johan Rönnby,
       editors, pp. 127–139. Southampton Archaeology
       Monographs New Studies No. 4. Highfield Press,
       Southampton.

LOEWEN, BRADLEY
2007   The Hull: of Ship Design and Carpentry. In
       *The Underwater Archaeology of Red Bay: Basque
       Shipbuilding and Whaling in the 16th Century.* Vol.
       III. R. Grenier, M. Bernier and W. Stevens, editors,
       pp. III-1–III-148. Ottawa.

LOPES, GONÇALO C.
2022   Boa Vista 1: estudo arqueológico de um navio
       na Lisboa Ribeirinha (séculos XVII-XVIII). (Boa
       Vista 1: An Archaeological Study of a Ship in
       the River Lisbon (17th–18th centuries). 2 Vols.
       Doctoral thesis, Faculdade de Ciências Sociais
       e Humanas da Universidade Nova de Lisboa.
       Available electronically from http://hdl.handle.
       net/10362/148186.

LOPES, GONÇALO C., JOSÉ BETTENCOURT, CRISTÓVÃO
FONSECA, TIAGO SILVA, INÊS P. COELHO, AND PATRÍCIA
CARAVALHO
2021   Early Modern Reused Ship Timbers from
       Boqueirão do Duro (Lisbon, Portugal). In
       *Proceedings of the 15th International Symposium
       on Boat and Ship Archaeology: Open Sea, Closed
       Sea - Local Traditions and Inter-regional Traditions
       in Shipbuilding*, Julia Boetto, Patrice Pomey, Pierre
       Poveda, editors, pp. 357–359, Marseilles, FR.

MATEUS, MARIANA
2018   O sítio do Boqueirão do Duro: contributo para
        o conhecimento da ribeira ocidental de Lisboa
        entre os séculos XVIII e XIX. (The Boqueirão do
        Duro site: A Contribution to the Knowledge of
        Lisbon's Western Riverfront between the 18th
        and 19th Centuries) Master's Thesis, Faculdade de
        Ciências Sociais e Humanas da Universidade Nova
        de Lisboa, Available electronically from http://hdl.
        handle.net/10362/63688.

MATEUS, MARIANA, RAQUEL SANTOS, NUNO NETO, JOSÉ,
BETTENCOURT, AND GONÇALO C. LOPES
2021   *Relatório preliminar – Rua da Boavista nº 51 a 59 e
        Rua D. Luís, nº 18, 18A e 18B, Lisboa. (Preliminary
        Report - Boavista Street, 51 a 59 and D. Luís Street,
        18, 18A and 18B, Lisbon.)* Neoépica, Lda., Mem-
        Martins.

McELVOGUE, D.
2009   Removing Water. In *Mary Rose: Your Noblest shippe:
        Anatomy of a Tudor Warship.* Vol. 2, P. Marsden,
        editor, pp. 288–295, Portsmouth, UK.

MENTZ, STEVE
2009   *At the Bottom of Shakespeare's Ocean.* Continuum
        International Publishing Group, London, UK.

MENTZ, STEVE
2015   *Shipwreck Modernity. Ecologies of Globalization,
        1550–1719.* University of Minnesota Press,
        London, UK.

OERTLING, THOMAS
1996   Ship's Bilge Pumps: a History of their
        Development, 1500-1900. College Station: Texas
        A&M University Press.

SCHOFIELD, JOHN, LYN BLACKMORE, AND J. E. PEARCE
2018   *London's Waterfront 1100–1666: Excavations in
        Thames Street. London. 1974–1984.* Archaeopress
        Publishing Ltd, Oxford, UK.

STANIFORTH, M.
1985   The Introduction and Use of Copper Sheathing: A
        History. *The Bulletin of the Australian Institute for
        Maritime Archaeology* 9:21–48.

THOMSEN, MIKKEL H.
2000   The Studland Bay Wreck, Dorset, UK: Hull
        Analysis. *International Journal of Nautical
        Archaeology* 29.1:69–85.

WADDELL, P.
1985   The Pump and Pump Well of a 16th Century
        Galleon. *International Journal of Nautical
        Archaeology* 14.3:243–259.

WHITEWRIGHT, JULIAN, AND JULIE SATCHELL
2011   *The Archaeology and History of the* Flower of
        Ugie, *Wrecked 1852 in the Eastern Solent.* British
        Archaeological Reports, British Series 551, Oxford,
        UK.

· · · · · · · · · · · · · ·

Mariana Mateus
FCSH-UNL/CHAM
Av. Berna, 26 C 1069-061 (Lisbon) - Colégio
Almada Negreiros (CAN), 330. Universidade
NOVA de Lisboa – Campus de Campolide
Lisbon, Portugal 1069-061

José Bettencourt
FCSH_UNL/CHAM
Av. Berna, 26 C 1069-061 (Lisbon) - Colégio
Almada Negreiros (CAN), 330. Universidade
NOVA de Lisboa – Campus de Campolide
Lisbon, Portugal 1069-061

Gonçalo Lopes
FCSH-UNL/CHAM; DGPC-CNANS
Av. Berna, 26 C 1069-061 (Lisbon) - Colégio
Almada Negreiros (CAN), 330. Universidade
NOVA de Lisboa – Campus de Campolide
Lisbon, Portugal 1069-061

Nuno Neto
Neoépica Lda
Rua do Rio, Quinta do Rebelo, Sacotes - Pavilhão
I, Algueirão - Mem Martins (Sintra)
Rio de Mouro, Mem Martins,
Sintra, Portugal 2725-524

Raquel Santos
Neoépica Lda
Rua do Rio, Quinta do Rebelo, Sacotes - Pavilhão
I, Algueirão - Mem Martins (Sintra)
Rio de Mouro, Mem Martins,
Sintra, Portugal 2725-524

Luís Reis
Neoépica Lda
Rua do Rio, Quinta do Rebelo, Sacotes - Pavilhão
I, Algueirão - Mem Martins (Sintra)
Rio de Mouro, Mem Martins,
Sintra (Portugal) 2725-524

Lisbon's Anchors: Archaeological Remains of its Maritime Past

Francisco Mendes, José Bettencourt, Marco Freitas

This paper presents the results of a study on iron anchors excavated since 1995 in archaeological contexts on the Lisbon waterfront. The collection comes from shallow subtidal or intertidal contexts related to port activities, where nautical structures, such as wharves, ships, and boats, have been documented. The research has revealed several typologies, with chronologies ranging between the 19th and 20th centuries. Their size is also diverse, pointing to their use on ships of widely varying tonnages.

Este artigo apresenta os resultados de um estudo sobre âncoras de ferro escavadas desde 1995 em contextos arqueológicos da frente ribeirinha de Lisboa. O espólio provém de contextos húmidos pouco profundos, relacionados com atividades portuárias, onde foram documentadas estruturas náuticas, como cais, navios e barcos. A investigação revelou várias tipologias, com cronologias que variam entre os séculos XIX e XX. A sua dimensão é também diversa, apontando para uma utilização em navios de tonelagem muito variada.

Cet article présente les résultats d'une étude sur les ancres de fer excavées depuis 1995 dans des contextes archéologiques dans le secteur riverain de Lisbonne. La collection provient de contextes subtidaux ou intertidaux peu profonds liés aux activités portuaires, où des structures nautiques, telles que des quais, des navires et des bateaux, ont été documentées. La recherche a révélé plusieurs typologies, avec des chronologies variant entre le 19e et le 20e siècle. Leur taille est également diversifiée, ce qui indique leur utilisation sur des navires de tonnages très variables.

Introduction

In recent decades, from the 1990s onwards, the archaeology of riverside areas has become an important field of work in Lisbon, Portugal. They have revealed a growing number of nautical structures buried under the landfills that have, in some cases, gained over 300 meters (m) of land to the river. These structures are varied, including waterfront revetments, shipbuilding structures, and a number of more-or-less structured remains of ships and boats, abandoned on the beach or in the shallow subtidal zone (Bettencourt et al. 2021). The same waterlogged contexts have also exposed important collections of objects related to maritime activity, which include pump tubes, net weights, cables, cannons, and iron anchors.

Anchors occupy a special place in maritime archaeology. They are a common element associated with port contexts, anchorages, or fishing traps, often related to the fishing of tuna, particularly common in the southern cost of Portugal (Baço 2014). They are a major icon of the nautical world—often alluded to in literature and folklore; represented in heraldry; used as decorative elements in buildings, murals, and paintings; and even tattooed on human skin as a marker of the owner's maritime identity. Their detailed study constitutes an important source of information on chronology, provenance, and even as an indicator of the capacity of ships that

used them on board (Ciarlo 2019:173–174). This paper aims to present a first approach to the modern anchors recovered in Lisbon waterfront in the context of urban requalification works.

General Context

This study includes the analysis of 14 anchors from three archaeological contexts excavated since the 1990s in western Lisbon, in the area that would roughly correspond to the old Santos beach. The first anchors were identified in 1995 at Cais do Sodré during the excavation of the underground tunnel. The context included a well-preserved ship, radiocarbon dated to the second half of the15th or the beginning of the 16th century (Rodrigues et al. 2001; Castro et al. 2011; Rodrigues 2020). The six anchors were identified in the sedimentary deposits that covered the ship, but no data is available on their context. These deposits also preserved pottery and other archaeological materials that have never been studied.

The second set was recovered during construction of the Promenade residential building, between Avenida 24 de Julho and Rua D. Luís, once again situated in western Lisbon. The archaeological work carried out allowed the identification of several wooden structures and deposits related to the waterfront, including about two dozen nautical timbers. Appearing mainly in fluvial sediments, these can be dated between the second half of the 18th century and the first half of the 19th century. Several pieces of iron nautical equipment were also located, including three anchors and two cannons (Fernandes et al. 2020).

Finally, the third group concerns the anchors recovered in 2020 during the construction of a new building, intended for a hotel and apartments, in the same area. The archaeological excavation revealed a long stratigraphic sequence, including a port area, located at depths between 2 and 7 m (above mean sea level [amsl]), featuring ceramics ranging from the Roman period to the 18th century. These contexts also preserved several pieces of nautical equipment, including iron anchors from between the 16th and 19th centuries, a small river boat (Boa Vista 4) and a well-preserved 17th-century ship. Above these levels, formed by silty sediments, several landfills were created, which were the basis for the construction of commercial and industrial buildings in the 19th century (Bettencourt et al. [2024]).

Methodology

The detailed study of anchor morphology is the best approach both to establish chronology and provenance, and to analyze the type of vessel that used them. The research, therefore, adopted a strategy of systematic documentation, with the goal of obtaining data that allows a reconstitution of the shapes of each anchor to scale. This documentation strategy included several phases of work. First, the pieces went through a cleaning process to remove sediments and the products of chemical and biological alterations that had accumulated on the surfaces. Documentation of the anchors was then carried out using several methodologies. Most pieces were scanned with an Artec Eva scanner. The pieces that were in limited spaces and had unfavorable environmental conditions, which limited the use of the scanner, were registered by Structure from Motion photogrammetry. Both methodologies made it possible to obtain high resolution 3D- or 2.5D-models. The analysis of these models was then performed in Rhinoceros software, which allowed the preparation of final prints and calculation of the volume of the pieces.

In the cataloging process, the authors adopted the proposal of the Association pour le Développement et la Recherche en Archéologie Maritime (Association for Development and Research in Maritime Archeology) or ADRAMAR, which includes parameters such as the position of the nut, the angle of the arms or the shape of the nails, in addition to 18 measurements considered

essential to comparative studies (Table 1). The measurements taken are described in Figure 1.

The weight was obtained by multiplying the volume calculated in Rhinoceros by the density of the cast iron—7.7 grams/cubic centimeters (g/cm^3) (Davis 1998 17). This methodology has limitations since the technology available between the 15th and 19th centuries, as well as the frequent use of raw materials of inferior quality, lead to imperfections in the manufacture and even hollow portions of the final product (Stelten 2010:32). Furthermore, one must not forget that post-depositional factors, such as concretion or the loss of parts of the anchor, can also affect the results obtained. For that reason, the volume was calculated only for complete or near complete anchors.

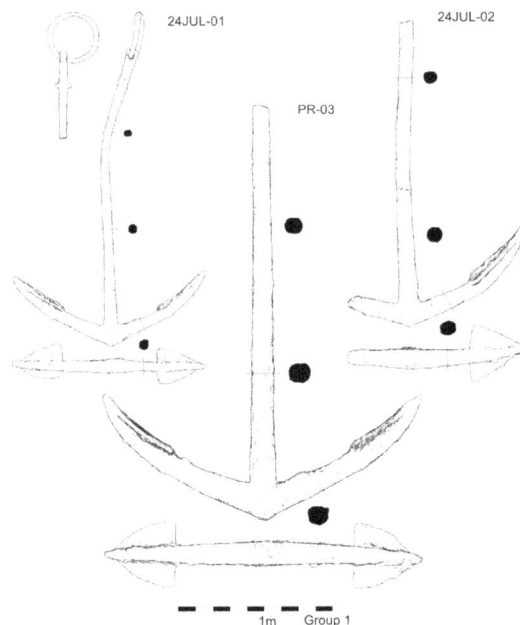

FIGURE 1. Technical drawing of anchor 24JUL-04 (including cross-sections), with the measurements taken during the recording of these objects (a) total height, (b) length of arms, (c) angle at the throat, (d) length of the head (e) Diameter of shank-end, (f) width at the throat, (g) length of one arm, (h) external diameter of the ring (i) internal diameter of the ring, (j) thickness of the ring, (k) thickness of the nuts, (l) length of the nuts, (m) width of the nuts, (n) length of the fluke, (o) width of the fluke, (p) thickness of the fluke, (q) length of the stock (not pictured) and (r) distance between iron bands on the stock (not pictured). Measurements presented in Table 1. (Figure by the primary author, 2022.)

I.D.	a	b	c	c'	d	e	f	g	h	i	j	k	k'	l	l'	m	n	o	p	q	r
PR-01	231	>133.0	87°	101°	51.0	6.6	12.7	80.6				3.7	4.0	1.6	1.9	11.0	35.7	36.6	1.5		
PR-02	277.5	109	82°		41.0	11.0	17.0	112.0				3.9	3.3	2.0	2.2	12.0	57.0	49.0	16.0		
PR-03	>244.0	184	102°	109°		9.6	15.0	117.0									51.0	38.0	15.6		
24JUL-01	160	104	105°	109°	34.0	3.8	7.0	68.0	27.0	22.5	3.0	2.0	2.5	2.0	2.1	8.5	33.3	20.0	4.6		
24JUL-02	>182.0	>105.0	101°	110°			11.0	89.0									48.0	31.0	7.0		
24JUL-03		129					10.0	66.3									31.0	26.0	3.5		
24JUL-04	190	104	93°	91°	38.0	4.5	8.5	62.0	23.6	18.0	3.5	2.9	3.5	2.4	2.7	4.2	30.0	31.0	3.5		
24JUL-05	251.5	152.5	101°	100°	39.5	6.7	11.7	76.0	40.6	29.8	4.6					11.2	44.5	36.0	2.1	2.7	
CS-01	>153.8	106	88°	87°			7.9	60.6									30.5	25.8	1.1		
CS-02	>165.5	93.7	100°	94°			5.7	56.7									31.8	>22.7	0.8		
CS-03																					
CS-04	>226.3				55.8	9.0			50.8	39.5	5.1	4.3	4.1	3.6	3.7	7.5					
CS-09																					
CS-014	306.9	193.4	106°	71°	63.8	10.2	13.8	112.6				4.2	3.1	3.7	3.2	7.8	55.6	51.0	2.9		

TABLE 1. Measurements taken to date of the anchors studied in this paper, detailed key of the measurements illustrated in Figure 1 cross-sections.

The data on the morphology of the anchors were then cross-checked with reference literature that explores formal variations in iron-anchor typologies during the modern era, based on specimens still underwater, sometimes in association with a shipwreck, and written and iconographic sources (Ciacchella 2021; Sadania 2009; Chouzenoux 2011; Curryer 1999). Some features may have chronological significance and contribute to the identification of production areas. The orientation of the nuts in relation to the arms of the anchor are particularly interesting and seems to have changed between the 16th and 17th centuries; during the 16th century, most anchors show the nuts on the same plane as the arms (Ciacchella 2021:128) and from the 17th century onwards, the nuts appear mostly on a plane perpendicular to the arms (Ciacchella 2021:129).

Other characteristics should also be considered, although these are much better studied in the case of French and British anchors. One can look at the shape of the arms, which, for instance, in the English case change from straight arms in the "Old Plan Long Shank" anchors, used from the mid-16th century until the early 19th century (Ciarlo 2019:175), to the more rounded shape of Pering's 1815 proposal (Curryer 1999:76).

The authors also took into consideration the morphology of the flukes, which mainly seems to be an indicator of provenance. This is a characteristic used as early as the mid-18th century (Deslongchamps 1759) to identify the nationality of a given anchor in an engraving showing the shape of anchors produced in various parts of Europe (Sadania 2009:49). Sometimes, when provenance is established, the proportion between the flukes and the rest of the arm can also provide information about chronology.

The chronology and provenience of an anchor can be further approached through the analysis of tables of proportions, existing since at least the 18th century for the British (Sutherland 1717, in Curryer 1999: 53; Steele 1794; Pering 1819, in Marlowe 2017:38–40) and French cases (Aubin 1702; Réaumur and de Monceau 1764; Diderot and D'Alambert 1762, in Sadania, 2009:61–64).

The tables regulating the quantity and type of anchors that each class of ship was to carry on board, also known at least since the 18th century for the British and French cases, are also an important source of information, allowing the establishment of roughly the size and type of ship that used an anchor found with no association to a shipwreck (Ciarlo 2019:174; Lescallier 1791, in Sadania 2009:67). Thus, the theoretical weight of each anchor calculated in Rhinoceros and the measurements of the shank and arms were the basis for trying to establish the anchor type, comparing this data with the values in the existing tables of proportions of the 18th and 19th centuries (for example, for the French case, Diderot and D'Allambert 1762, in Sadania 2009:61–62; Réaumur and de Monceau 1764, in Sadania 2009:63; Aubin 1702, in Sadania 2009: 64; Lescallier 1791, in Sadania 2009: 67). This exercise remains in the realm of hypotheses because these regulations would primarily apply to national navies and did not necessarily have equivalents for all other European nations. Furthermore, the results obtained through these tables should be considered with reservations, because, for example, an anchor playing the role of a bow anchor aboard a large ship may correspond in weight to the stream anchor aboard a smaller ship.

The Anchors

The comprised 14 anchors are in different states of preservation. The anchors (24JUL-02, PR-01, PR-02, PR-03, CS-01, CS-02 and CS-04) are incomplete and damaged, which prevents a classification with maximal reliability (Table 1). In one case (24JUL-03), the authors were unable to obtain any useful data as only the arms, with both flukes attached, remained, but in too poor condition to make out the original shape. Anchor CS-04 is only composed of a portion of its shank and the ring, still attached. The other anchors are in better general shape, although some elements may be missing, such as the ring or a fluke.

Looking at the general morphology, the 14 anchors can be divided into two main groups: the anchors whose arm shape presents an arch separated into two relatively straight sections, with the second one at a steeper angle, and creating a pointed crown under the shank (24JUL-01 and 02, PR-03, CS-03, CS-09) as illustrated by a selection of artefacts in Figure 2; and the anchors with curved arms, corresponding to the shape that is more traditionally associated with an anchor in popular imagery, forming part of a circle that joins the shank roughly in its center (24JUL-04 and 05, CS-01, CS-02) shown in Figures 1 and 3. Three anchors do not fall into these categories, marked by the wider angle at which the arms are welded to their shanks (PR-01, PR-02, CS-14) (Figure 4).

Group 1 - The French, or French-style, Anchors

Based on the bibliography and historical iconography, the anchors in the first group may have French origins or may have been produced following late 17th- and 18th-century French practice, in the shape of the anchor reproduced in Diderot and D'Alambert's *Forge des Ancres* (Chouzenoux 2011:80). Beyond the similarity in shape of the arms, in the case of anchors 24JUL-01 and

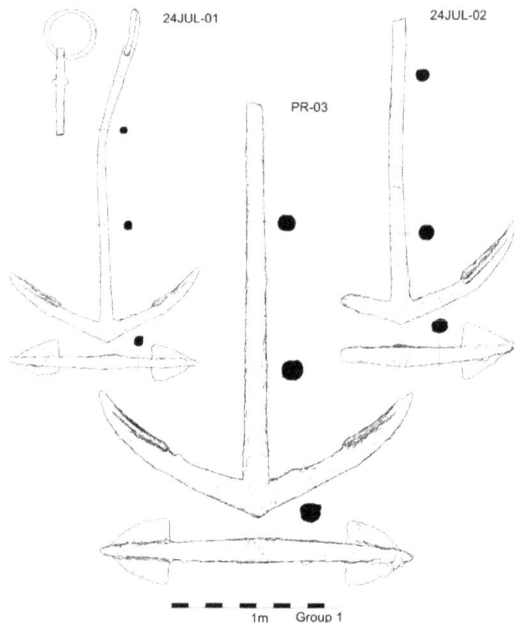

FIGURE 3. Technical drawing of the anchors in the second group including the stock of 24JUL-05. (Figure by the primary author, 2022.)

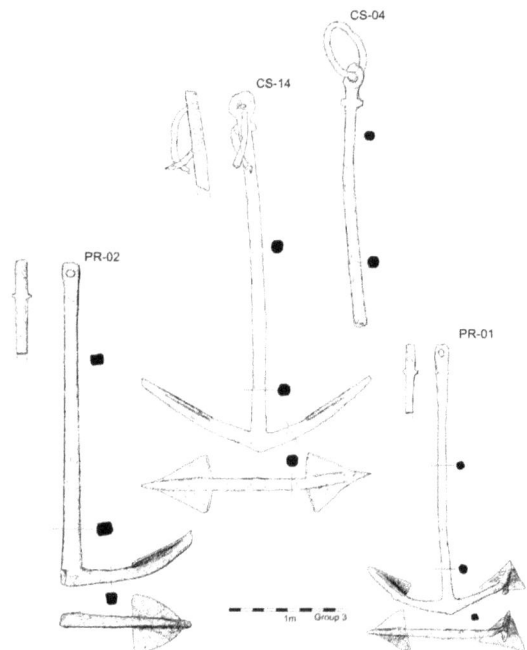

FIGURE 2. Technical drawing of selected anchors from the first group. (Figure by the author, 2022.)

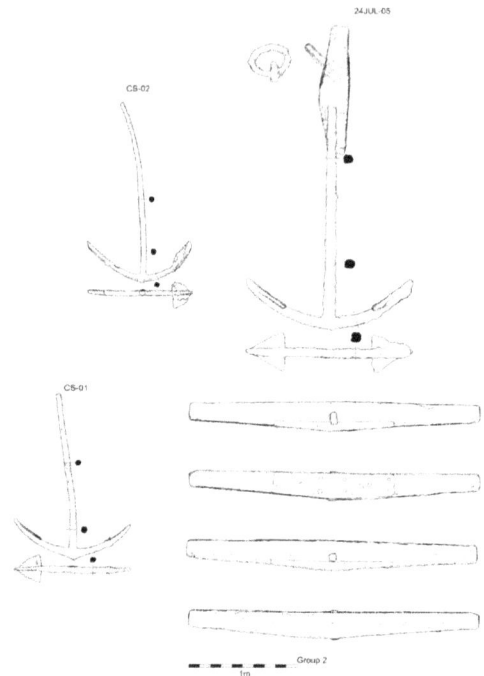

FIGURE 4. Technical drawing of the anchors in the third group, with the addition of anchor CS-04, whose original shape is unknowable. (Figure by the primary author, 2022.)

24JUL-02, the shape of the flukes is also identical to that commonly associated with the Toulon dockyards (Sadania 2009:55–56).

The riveting of the legs to the anchor arm, another feature associated with anchors produced in the Toulon dockyards during the 18th century (Deslongchamps 1759, in Sadania 2009:55–56), is more difficult to ascertain because the 24JUL-02 flukes are deformed by concretion. However, in the case of 24JUL-01, no evidence has been identified that suggests the presence of a rivet reinforcing the connection of one of the flukes to the arm. This does not appear to be repeated on the opposite fluke and may, therefore, be evidence of repair. This is only one of the possible scenarios, since Deslongchamps drew in 1732 a similar anchor where only one of the flukes appears to be riveted (Sadania 2009:56). However, doubt will remain: is it an option made during the forging process, the representation of a repair, or merely the result of the artisan's choice? It should also be noted that the arm perforation would always constitute the introduction of a weak point in the structure of the anchor, something that could be seen as a greater disadvantage than the gains in the strength of the connection between the legs and the arm (Sadania 2009:55). The repair hypothesis is the most probable scenario.

Anchors PR-03, CS-03, and CS-09 have flukes of a different design, more closely resembling the shape of a spade than a triangle, with the sides of the flukes being slightly rounded. According to Deslongchamps' illustrations of 1759 (Sadania 2009:58–59), this fluke shape would correspond to types produced in Holland, Denmark, or Spain. However, the first plate of *Fabrique des Ancres* by Réaumur and Monceau (1764) presents a fluke shape closely resembling that of this group, representing what would be a typical French anchor during the 1760s (Ciarlo 2019:177). This presents a completely different morphology from that represented by Diderot and D'Alembert in their *Forge des Ancres* published in the same decade (Ciarlo 2019:178). In the archaeological record, similar features in anchors have been recorded in the bay of Angra do Heroísmo, in the Azorean Terceira Island, also identified as being of French origin (Chouzenoux 2011:79–82).

The shank cross-section of all these anchors appears to be the same, describing roughly a hexagonal shape, eroded to different degrees depending on the piece. This is a shape that seems to be less common in the archaeological record, because of the 11 anchors identified in Angra do Heroísmo as being of French origin, only one

has this type of cross-section (Chouzenoux 2011:86), a fact made more interesting by the many graphic representations of French anchors of the period as having six-sided shanks (Réaumur and Monceau 1764, in Ciarlo 2019:177; Diderot and D'Alambert 1762, in Sadania 2009:46; Deslongchamps 1731, in Sadania 2009:51; Vial de Clairebois 1787, in Ciacchella 2021:122; Lescallier 1791, in Ciacchella 2021:122). The 24JUL-01 anchor is perhaps the most interesting, as it is the one with the thinnest shank, giving it an appearance more easily associated with the anchors of Iberian tradition (Chouzenoux 2011 87). The angle of the arms at the junction with the shank is also similar within the group, having a maximal variation of four degrees (between 101 degrees of 24JUL-02 and 105 degrees of 24JUL-01) among the 3 anchors that were measured.

Unfortunately, only two of the anchors in this group have their heads intact, anchors 24JUL-01 and CS-03. In both cases, it is of sub-quadrangular section and ends in a rectangular shape. The nuts are on a plane perpendicular to the arms, the most common configuration since the 17th century and until the end of the use of wooden stocks (Ciachela 2021:129).

The comparative analysis of these anchors, between the size and weights, and the French tables of proportion allowed the authors to propose the different ships they may have equipped. Anchor 24JUL-01, missing only it's stock, would have a weight of 79.57 kilograms (kg), exceptionally low for the set weighed due to its comparatively thin shank. The length of the arms and some other proportions, such as the measurements of the flukes and of the head, are in line with an anchor of 200 pounds (lbs.) (c. 97.90 kg, assuming that these are the old Paris pounds), according to Reaumur and de Monceau (1764, in Sadania 2009:63), though in reality it has 20 kg less than its theoretical weight.

Anchor 24JUL-02 is incomplete, weighing 155.53 kg. Considering Réaumur's table, it would correspond to an anchor of about 600 lbs. (Sadania 2009:63), heavy enough to appear in Lescallier's table as being a 1st or 2nd streamer anchor aboard a corvette. According to Aubin, this would be an anchor suitable for a ship with a beam measuring 21 *pieds* (6.82 m), although, given that this table is older, the proportions between weight and length of the shaft vary; for this same type of anchor, Diderot mentions a theoretical length of 2.22 m (6 *pieds*, 10 *pouces*) and Aubin of 2.76 m (8 *pieds*, 6 *pouces*) (Sadania 2009: 61, 64).

Anchor PR-03 has a real weight of 434.75 kg corresponding in theory to a 1000 lb. anchor (489.5 kg), since it is missing its head and ring. These anchors would have a length of 3.14 m (nine *pieds*, eight *pouces*) according to Diderot (Sadania 2009:61) and 10 *pieds* according to Aubin, a discrepancy of only 10 cm (four *pouces*). This was suitable to equip a ship with a beam of 25 *pieds* (8.12 m) according to Aubin (1702), or to be the first streamer aboard an eight-gun frigate according to Lescallier (1791) (Sadania 2009:64, 67).

Group 2 - The Anchors with Curved Arms

The second group consists of anchors in which the shape of the arms is curved, forming a portion of a circle, or an arc. This shape goes back to antiquity, documented for example in the ships of Lake Nemi, in Italy, dated c. 40 AD; it would also become the most common anchor design in the 16th century (Ciacchella 2021:112) and would gain popularity again during the 19th century following the innovations brought about by Richard Pering (Curryer 1999:76). This group is in a relatively better state of conservation than the previous one. This allows observation of almost all the characteristics that can lead to the identification of the chronological and cultural signatures anchors can contain, except for CS-01 and CS-02, of similar shape to the others, but too small and incomplete to offer any conclusions. This leaves anchors 24JUL-04 and 24JUL-05, the latter being the only one still with its wooden stock intact.

Anchor 24JUL-04 seems to be the oldest of the whole collection, since its nuts are in the same axis as the arms. Its chronology can be further refined by considering the shape of its head (Ciachella 2021:128). Unfortunately, the presence of concretion around the head-eye and the ring makes it difficult to analyze. Some evidence of slight bulging is present, placing this artifact somewhere between the late 15th century and the first quarter of the 17th century (Ciachella 2021:128). Two other anchors possibly from the same period, CS-04 and CS-14, are discussed below. These features, together with the triangular flukes welded to the curved arms of the anchor, represent the most common features in 16th-century anchors (Chiacchella 2021:112). The date of anchor 24JUL-05 is more difficult to establish, as the nuts are hidden by the wooden stock. However, its general shape appears identical to 24JUL-04, differing only in dimensions (24JUL-05 is 50-cm taller), but retaining its chronological signatures and proportions.

The wooden stock of 24JUL-05 is of a model rarely represented by historiography, in part because objects made of perishable material rarely survive. The closest examples to this stock come from Northern Europe

(Votruba 2022:340; Cicchella 2021:109). In these cases, a secondary timber of the stock was fixed around the anchor to a larger piece of the stock previously shaped with a mortise like its shank and nuts. This model differs from the most common type of wooden stock, composed of two symmetrical halves, fixed together by nails or treenails, and reinforced by iron hoops (Sadania 2009:40–41). The section of the shanks is very thin, of sub-quadrangular shape, and the arms are relatively short. In both cases, one arm represents a third of the length of the shank, between the throat and the beginning of the head square. One could apply to all the examples in this group the Dutch nautical expression "thin as a Spanish anchor" (Curryer 1999:44), sharing between them characteristics that would be typical of anchors produced in the Iberian Peninsula before the 17th century (Chouzenoux 2011:87). This is a time and place for which, unfortunately, tables of proportions, such as those applied to the anchors of the previous group, are not available.

Group 3 - The "Open" Anchors

Finally, the last type of anchor is characterized by its more open shape at the arms, a shape caused by a wider angle between the shank and arms. Both PR-01 and PR-02 are damaged; the former has one of its arms broken and twisted laterally, with the fluke still attached to the end that has moved, and the latter has lost one of its arms completely.

Anchor PR-02 is similar to one of the examples studied in Angra do Heroísmo (Chouzenoux 2011:91), which the author describes as comparable to models used by the German, Dutch, and Swedish navies around the year 1800. A Northern European origin is also suggested by a drawing from 1859 that shows different anchors in use at the time, where an anchor of similar shape is identified as a model in use by the Dutch, Danish and Swedes (Cotsell 1856 in Stelten 2010:33). The shape of the flukes is also represented in the Dutch anchors on the plate "*Ancres de divers pays*" (Deslongchamps 1759, in Sadania 2009:54). The lack of PR-02's arm does not rule out the hypothesis of it being a mooring anchor designed for continuous use in anchorages, where a second arm sticking out of the sand could prove dangerous to ships using said anchorage (Sadania 2009:77). However, upon close observation, the authors are more inclined towards the possibility of it being a regular anchor that lost an arm in use, possibly reused as a mooring anchor, a practice previously documented (Sadania 2009:78–79).

Anchor PR-01 shows a similar arm shape to PR-02, but has a more pronounced angle at its second segment, which gives it a slightly more closed appearance. This shape is reminiscent of some French anchors (Chouzenoux 2011:84) in which the shape of the arms, being neither curved nor straight, forms part of an octagon (Chouzenoux 2011:83). This would possibly be another example of a French anchor, although in this case the shape of the flukes is not the one most associated with that origin. However, the cross-section of the shank is the same as seen on the French anchors above and the weld of the flukes to the arms were both reinforced with a rivet. This is probably the reason explaining the abandonment of this artifact, because one of the holes opened to receive one of the rivets appears to be the weak point where the arm gave way, perhaps confirming Brest shipyards' suspicions of this kind of reinforcement in the 1730s (Sadania 2009:55).

Finally, anchor CS-14 likely represents a 16th century anchor of unknown origin. It shares the same features mentioned above for 24JUL-04 and 24JUL-05, which it would have also likely shared with most anchors of its period (Chiacchella 2021:112), being only different in the shape of the arms. It is worth mentioning that anchor CS-04 features a shank with an identical head, probably sharing with CS-14 the chronology and provenance.

Final Remarks

The anchors found on the Lisbon waterfront represent several typologies used throughout the modern era. The oldest (CS-04 and CS-14) likely correspond to models from the 16th century, sharing most or all the characteristics associated with those anchors: nuts on the same plane as the arms, triangular flukes, and bulging of the head clearly showing gothic finials (Chiacchella 2021:128).

The most recent ones appear to date to the 18th century and to be mostly of French origin, having details that appear on Réaumur's *Fabrique des Ancres*, such as the flukes in the shape of a spade found on PR-03, CS-03, and CS-09, and other details drawn by Diderot and D'Alembert's on the *Forge des Ancres*, like the shape of the arms, identified as French, or of French style.

By looking at the size and weight, one can assume that Lisbon anchors studied would be in use aboard small- to medium-sized ships. For those thought to be of French origin, the authors were able to compare the data obtained during the recording process with the tables produced during the 18th century. According to these records, an anchor such as 24JUL-02 could be suitable for the operation of a ship, like a corvette; anchors such as PR-03, larger in size and weight, could equip a ship like a frigate.

The present research corresponds to the beginning of a systematic study of the anchors found in the Lisbon riverfront, integrated in a project of much wider scope on the maritime archaeology of Lisbon and the Tagus. This first analysis allowed the authors to verify that this study on anchors can contribute significantly to furthering knowledge about the maritime dimension of the port, providing clues about the dynamics of navigation in place during the modern period. The continuity of this line of research is therefore justified.

Acknowledgements

This paper had the support of CHAM (NOVA FCSH/UAc), through the strategic project sponsored by FCT (UIDB/04666/2020). We would like to thank Catarina Rosa for reviewing the English version.

References

AUBIN, NICOLAS
1702   Dictionnaire de Marine (Nautical Dictionary). Amsterdam, NL.

BAÇO, JOANA
2014   Âncoras ao Largo: um Contributo Arqueológico para o Estudo das Atividades Marítimas em Lagos na Idade Moderna (Anchors at Large: The Archeological Contribution for the Study of Maritime Activities in Lagos during Post-Medieval Times), Master's thesis, Faculdade de Ciências Sociais e Humanas, Universidade Nova de Lisboa, Lisbon.

BETTENCOURT, JOSÉ, C. FONSECA, P. CARVALHO, T. SILVA, I. COELHO, AND G. LOPES
2021   Early Modern-Age Ships and Ship Finds in Lisbon's Riverfront: Research Perspectives. *Journal of Maritime Archaeology* 16(2):1–31.

BETTENCOURT, JOSÉ, P. CARVALHO, M. PONCE, T. NUNES, G. LOPES, T. SILVA, AND I. MENDES DA SILVA
[2024]   An Extraordinary Find? Boa Vista 5, A New Early Modern Ship Discovered in Lisbon Waterfront. Paper presented at the 16th International Symposium on Boat & Ship Archaeology, Zadar, HR.

CASTRO, FILIPE, K. YAMAFUNE, C. EGINTON, AND T. DERRYBERRY
2011   The Cais do Sodré Shipwreck, Lisbon, Portugal. The *International Journal of Nautical Archaeology* 40(2):328–343.

CHOUZENOUX, CHRISTELLE
2011   Charactérisation et Typologie du Cimetière des Ancres (Characterisation and Typification of an Anchor Cemetery), Master's thesis, Universidade Fernando Pessoa, Oporto.

CIACCHELLA, FABRIZIO
2021   The Evolution of Nuts and Other Fixation Systems of Wooden Stocks to Iron Anchors. *Archeologia Postmedievale* 25:107–131.

CIARLO, NICOLÁS
2019   Aportaciones Históricas y Arqueológicas al Estudio del Proceso de Estandarización en la Industria Ancorera de las Potencias marítimas Europeas del Siglo XVIII (Historical and Archaeological Contributions to the Study of the Standardization of the Anchor-Producing Industries of the European Maritime Powers of the XVIII Century). *Arqueología* 25(2):169–193.

COTSELL, GEORGE
1856   *A Treatise on the Ship's Anchors.* London, UK.

CURRYER, BETTY
1999   *Anchors: An Illustrated History.* Chatham Publishing, London, UK.

DAVIS, JOSEPH R. (EDITOR)
1998   *ASM International Metals Handbook Desk Edition.* Novelty, OH.

DESLONGCHAMPS, L'AINÉ
1731   *Plan d'une Ancre du Poids de 2680 Livres (Plan of a 2680 lb. Anchor).* Cosne, FR.

DESLONGCHAMPS, L'AINÉ
1759   *Ancres de Divers Pays (Anchors of Varied Countries).* Brest, FR.

DIDEROT, DENIS, AND J. R. D'ALAMBERT
1762   *Forge aux Ancres in Encyclopédie ou Dictionnaire Raisonné des Sciences, des Arts et des Métiers: Marine, Recueil de Planches sur les Sciences, les Arts Libéraux, et les Arts Méchaniques, avec leur Explication (Anchor's Forge Encyclopedia).* Paris, FR.

FERNANDES, FÁBIO, A. DE ALBUQUERQUE, AND J. BETTENCOURT
2020   *Edifício Promenade, Avenida 24 de Julho nº10 e Rua D. Luís nº7, Lisboa Relatório Preliminar (Preliminary Report on the 2020 Promenade Building Excavations of 2020).* From Arqueologia e Património Lda., Empatia-Arqueologia, Lda. and Império Arqueologia.

LESCALLIER, DANIEL
1791   *Traité du Gréement des Vaisseaux et autres Batiments de Mer (Treatise on the Rigging of Ships and Other Naval Structures).* Paris, FR.

MARLOWE, ELIZABETH
2017   "The Bower yet Remains": Historical and Archaeological Technomic Analysis of Anchor Design in the Long Nineteenth Century, Master's thesis, Faculty of the department of history of East Carolina University, NC.

PERING, RICHARD
1819   *A Treatise on the Anchor.* Plymouth, UK.

RÉAUMUR, R. A.FERCHAULT, AND H. L. DUHAMEL DE MONCEAU
1764   *Fabrique des Ancres, lue á l'Académie en Juillet 1723, par M. de Réaumur avec des Notes et des Additions de M. Duhamel (Anchor Factory, read at the French Academy in 1723 by M. Réaumur and complemented by M. Duhamel).* Paris, FR.

RODRIGUES, P.
2020   O Estudo do Cavername do navio do Cais do Sodré da 2.ª Metade do Século XV/ Inícios do Século XVI (A Study of the Framing Timbers of the Cais do Sodré Ship, from the Second Half of the 15th Century/16th Century). Trabalhos de Arqueologia 55.

RODRIGUES P., F. ALVES, E. RIETH, AND F. CASTRO
2001   L'Épave d'un Navire de la Seconde Moitié du XV.ème Siècle/Début du XVI.ème, trouvée au Cais do Sodré (The Remains of a Ship from the Second Half of the 15th Century/16th Century found in Cais do Sodré, Lisbon) (Lisbonne). *Trabalhos de Arqueologia* 18: 347–380.

SADANIA, MARINE
2009   Les Ancres en Fer en Bretagne (Iron Anchors in the Bretagne Region of France), Master's thesis, Université de Rennes, Rennes, FR.

STEELE, DAVID
1794   *Elements and Practice of Rigging and Seamanship.* London, UK.

STELTEN, RUUD
2010   Relics of a Forgotten Colony: The Cannon and Anchors of St. Eustatius, Master's thesis, Faculty of Archaeology, Leiden University, BE.

SUTHERLAND, WILLIAM
1717   *England's Glory or Ship-Building Unvail'd.* London, UK.

VIAL DE CLAIREBOIS, HONORÉ-SÉBASTIEN
1787   *Encyclopédie Méthodique de la Marine (Methodical Encyclopedia of the Navy),* vol.4, Plates. Paris, FR.

Votruba, Gregory
2022   Iron-Frame Wooden Stock-Anchor Design in
        the Age of Exploration. *International Journal of
        Nautical Archaeology* 51(2):338–357.

. . . .

Francisco Mendes
CHAM - Centre for the Humanities
and History Department, FCSH
Universidade NOVA de Lisboa
Lisbon, Portugal

Marco Freitas
CHAM - Centre for the Humanities
and History Department, FCSH
Universidade NOVA de Lisboa
Lisbon, Portugal

José Bettencourt
CHAM - Centre for the Humanities
and History Department, FCSH
Universidade NOVA de Lisboa
Lisbon, Portugal

Gran Principessa Di Toscana: The Story and Archaeology of a 17th-Century Shipwreck in Cabo Raso (Cascais)

Sofia Simões Pereira

In the 1960s, a set of bronze cannons was discovered by an amateur diver in the Cabo Raso area of Cascais. In the years that followed, the archaeological site was the target of recoveries made by amateur divers. It was only in the 1980s that pioneering underwater archaeological work was carried out on this site. More recently, historical research identified the wreck as the Gran Principessa di Toscana, *which sank in 1696. This study focuses on the finds and documentation available about the wreck, with an attempt to deepen knowledge about the underwater archaeological heritage of Cascais.*

Na década de 1960, um mergulhador amador descobriu um conjunto de canhões de bronze na zona do Cabo Raso, em Cascais. Nos anos que se seguiram, o sítio arqueológico foi alvo de recuperações efetuadas por mergulhadores amadores. Só nos anos 80 é que foram efetuados trabalhos pioneiros de arqueologia subaquática neste local. Mais recentemente, a investigação histórica identificou o naufrágio como sendo a Gran Principessa di Toscana, *que naufragou em 1696. O presente estudo incide sobre os achados e a documentação disponível sobre o naufrágio, procurando aprofundar o conhecimento sobre o património arqueológico subaquático de Cascais.*

Dans les années 1960, un ensemble de canons en bronze a été découvert par un plongeur amateur dans la région de Cabo Raso à Cascais. Dans les années qui ont suivi, le site archéologique a été la cible de récupérations effectuées par des plongeurs amateurs. Ce n'est que dans les années 1980 que des travaux archéologiques subaquatiques pionniers ont été effectués sur ce site. Plus récemment, des recherches historiques ont identifié l'épave comme étant le Gran Principessa di Toscana, *qui a coulé en 1696. Cette étude se concentre sur les découvertes et la documentation disponibles sur l'épave, avec une tentative d'approfondir les connaissances sur le patrimoine archéologique subaquatique de Cascais.*

Introduction - Site History

The number of historical shipwrecks at the entrance of the Tagus River recorded since at least the 16th century amounts to about 100. Sources indicate a great cultural diversity, with records relating to Portuguese, Spanish, English, French, and Germanic ships, lost on a point of strategic importance for navigation to Lisbon, and a passage for ships heading to Northern Europe and the Mediterranean (Borges 2014:152; Freire and Fialho 2013a:1214).

In recent decades, direct evidence of ships arriving in Lisbon have been revealed, corresponding to shipwrecks and materials recovered in anchorages and landing areas through fortuitous finds and archaeological research. Most of the sites where cannons or anchors predominate, do not provide accurate information to their date, function, area of operation, or ship identification. Among these are the sites Baixa da Boeira, with eight cannons; Praia de Carcavelos Beach, with 3 anchors and 12 cannons; and Porto de Recreio de Oeiras, with 5 cannons. The only exceptions of the early-modern age are the *Nossa Senhora dos Mártires* shipwreck, in São Julião da Barra, Oeiras, lost in 1606 when returning from India, and the 17th-century Cabo Raso shipwreck (Bettencourt, et al. 2018:153).

The Cabo Raso shipwreck is among the first sites to be identified in Portugal. Its discovery dates to the 1960s, recorded by letter of 7 June 1966, from the Portuguese Federation of Underwater Activities (FPAS), which reports that on the previous day an amateur diver discovered a set of bronze cannons in an area near Cabo Raso. In the same letter, the FPAS requested authorization to proceed with preliminary documentation of the finds (drawing, exact location, photographs, etc.). It was also requested that protective measures be taken over the artifacts, due to their exceptional conditions (Cardoso 2012:8; Freire and Fialho 2013b:26; File 1966/01).

However, during the following decades (the 1960s and 1970s), the site would be the target of constant looting, "well-intentioned" recoveries made by amateur divers, and occasional retrievals by fishing boats. It is the case of the sailor José dos Reis Jorge (owner of the boat named "Praia da Nazaré"), who on 18 June 1972, at the request of the Captain of the Port of Cascais, recovered from the bottom of the sea, near the Lighthouse of Cabo Raso, a bronze Florentine culverin, with a coat-of-arms thought to be from the Medici. The discovery of this piece was the subject of several newspaper articles, which reported the circumstances of the discovery, the deciphering of the inscriptions of the cannon, and the discussion of its uncertain future (Atanásio 1981/1982:9; Cardoso 2012:9; Freire and Fialho, 2013b:27; File 1966/01).

From the 1980s onwards, the Museu do Mar de Cascais (Cascais Sea Museum) was able to carry out some work at the archaeological site—including the recording and positioning of some cannons, and the recovery of some materials, such as the second culverin on 10 May 1980. On 24 March 1998, João Pedro Cardoso made a routine dive, aiming to take measurements of a cannon, and found six more in its vicinity. In 2013, a new archaeological investigation was carried out at Cabo Raso, which aimed to monitor and record the evolution of the site (Cardoso 2012:10; Freire and Fialho 2013b:27–30).

The possible identification of the wreck was made in 1990 by Patrick Lizé, who located three documents related to the wreck of Italian, English, and French origin. In the English document, dated 1696, Paul Methuen mentions that the English mail failed again that week, due to bad weather. He also mentions that a large ship called the *Gran Principessa Di Toscana*, with 70 cannons on board, washed ashore about eight days before on the Lisbon rock, being lost with 60 men (Cardoso 2012). The French document dated 23 January 1697, states that the ship *Gran Principessa Di Toscana*, commanded by Captain Benoict Prasca of Livorno, had sunk at Cabo da Roca on St. Andrew's Day at midnight after three days of storms, a league away from Cascais, and most of the crew drowned. It mentions that the ship's crew knocked down all the masts and they deployed three anchors to the sea, which still did not prevent the ship from sinking (Cardoso 2012) This is justified by a letter dated 16 December 1696, written by Mr. Gautier a passenger saved on the rigging of the mizzen (which came ashore), addressed to Mr. Charles Ollivier, merchant in the street Bonnetterie in Marseille (Cardoso 2012:11).

These sources provide some of the ship's attributes, its origin, the name of the captain, the approximate date of the tragedy, the circumstances, and the approximate location. However, the finds recovered in the past are scattered in public and private collections and have never been the subject of detailed study. The only publication available is a catalog published by João Pedro Cardoso, a pioneer in underwater archeology in Portugal (Cardoso 2012). This paper presents a preliminary analysis of the site and known recovered artifacts. It is part of the author's Master's degree research, and it is framed within the activities of the project Early Modern Shipwrecks under Tagus Mouth - SUNK.

The Site and the Current Investigation

The Cabo Raso wreck is in the municipality of Cascais, a district of Lisbon, between two prominent points on the north coast at the mouth of the Tagus River, Cabo da Roca and the Fortress of São Julião da Barra (Figure 1). It is in a maritime area designated as Enseada Entre os Cabos (Between Cables Cove) because it is limited by Cabo da Roca, to the north and Cabo Espichel, to the south, and cut by the Tagus River (Freire et al. 2012:1).

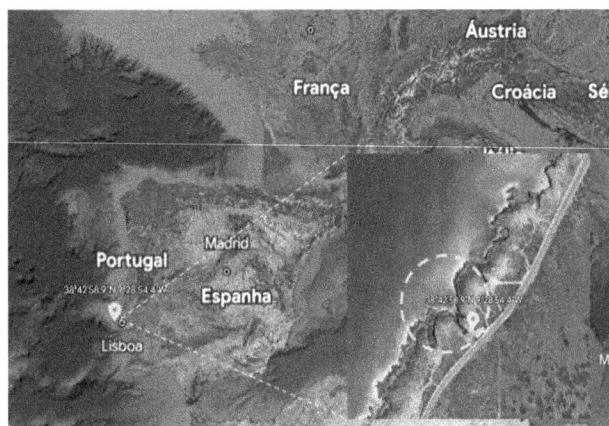

FIGURE 1. Location of the Cabo Raso site. (Image from Google Earth, <https://earth.google.com/>. Accessed 06 January 2023.)

The available data about the wreck and investigations at the site are scarce. In 1993, João Pedro Cardoso made the first survey of the cannons. In 2013, the Project of the Underwater Archaeological Chart of Cascais - ProCASC (Projeto da Carta Arqueológica Subaquática de Cascais) mission conducted a new survey on the site. The archaeological work was done in draft form, focusing on the collection of measurements, orientation, and depths of the cannons. The two documents allow a preliminary analysis of the archaeological site. The Cabo Raso site presents a west-east orientation, consisting of a set of cannons and a small concentration with an unidentified iron cluster. The topography of the site is variable. The artifacts are scattered between two depressions on the rocky bottom, on a long and wide re-entrance in the coast, which gives it a "bay" appearance. It is also possible to see the hydrodynamics of the landscape, which resembles a centrifugation site. The first kiddle

32   Advisory Council on Underwater Archaeology

(in Portuguese *caneiro*, an arm of the sea between rocks) facing west has a depth of about 5 meters (m), preserving most of the cannons in situ, and is interrupted in the middle by a rocky outcrop. North of the elevation is an inclined slab that goes from 5.7 m to 2.0 m in depth. This gives access to the second kiddle, which does not show great sedimentation, leading to less retention of artifacts. Outside the "bay", there is strong sedimentation (Freire and Fialho 2013b:30–31). In total, 16 cannons were in situ, 15 of iron, and 1 of bronze (two more than in 1993).

The Artifacts

Cabo Raso artifacts are diverse, including a collection of a bronze ordinance, an important source of information about the wreck, which in this case was essential to identifying the ship. The collection is composed of 19 pieces classified into two distinct groups: the Culverin and the Breech Loading Swivel Guns (Alves 1994:132) (Figure 2).

The culverin is a type of cannon with a relatively long barrel in relation to the caliber of the piece and consequently has a longer range. These pieces are loaded through the mouth (Salgado 2002–2006). One of the bronze culverin, with a total length of 2 m, was cast in Florence. It has two Latin inscriptions, from which the name of the individual who cast it, the name of the Prince, and the year of casting are present. The first inscription, carved above the second reinforcement on a cartouche with an eight-pointed cross, is FER (DINANDO) II HERTR (VRIAE) // MAG (NO) DVCE // MDCXXXXVII, which reads "Ferdinand the Second, Fifth Grand Duke of Etruria, 1647." The second inscription appears on the belt of the first reinforcement of the culverin, on the right side of the ear: OP (VS) IO (AN) NIS MARIANE // CENNII FLOREN (TI) NI. When translated, it reads "Work of John Mary Cenni, Florentine (gunsmith)." The second culverin recovered in 1980 is similar to the first one, but is eroded from sea action, which makes reading difficult (Atanásio 1981/1982:14; Cardoso 2012:23).

The other type of artillery found in the archaeological site are 17 breech-loading swivel guns, of which 11 are in private collections. These small arms, with a total length of 1.06 m and of small caliber, are loaded with mug-shaped chambers in which the gunpowder and projectile had been filled in advance (Alves 1994:132; Ridella, et al. 2016:186). In one of these guns, the mark of Amsterdam, with three XXX, is noted, which serves as proof of quality. A breech-loading gun and two chambers similar to the ones at Cabo Raso (the cannon also has the Amsterdam mark) were identified on the wreck of the *Slot ter Hooge,* lost in Porto Santo, Madeira Islands, in 1724 (Esmeraldo 2022). Also related to armaments recovered at the site are two muzzle guns (one shows the mark of Amsterdam) and five stone projectiles (round shot, one big and four of small caliber), half musket balls, and several lead bullets.

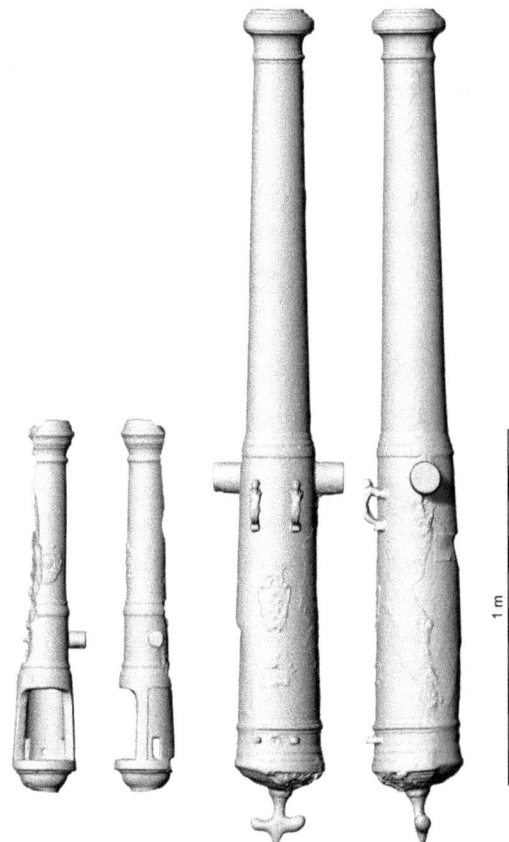

FIGURE 2. Three-dimensional model of the culverin (0032.04.01) and the breech-loading swivel gun (0032.04.03). (Figure created by the author, 2023.)

In addition to the military equipment, other materials have also been recovered, including day-to-day items of the ship, like one brass barrel tap, two bronze bells, two copper candlesticks, and six pewter plates. The plates confirm the shipwreck date because they present the broad-rimmed type (plain without decoration). Three of them present different manufacturing marks, which have not yet been identified. One bronze frame, decorated in relief with phytomorphic motifs and two figures of angels (Figure 3), has also been identified.

Nautical equipment includes a depth sounding in lead, octagonal in section, and truncopyramidal in shape. A copper compass is incomplete, only preserving the upper part and part of one of the legs (Figure 3). The legs would be pointed at their lower ends, and the upper ends form two concentric arches joined in a hinge. An aspect of this type of compass is that the two arches that form the top of the instrument open into double slats and embrace each other. This process allows the instrument to be adjusted to the desired opening and transported to a new point on the chart. This is a common model from the 17th century; three other examples were recovered in São Julião da Barra (Directorate-General for Cultural Heritage (DGPC) 2010).

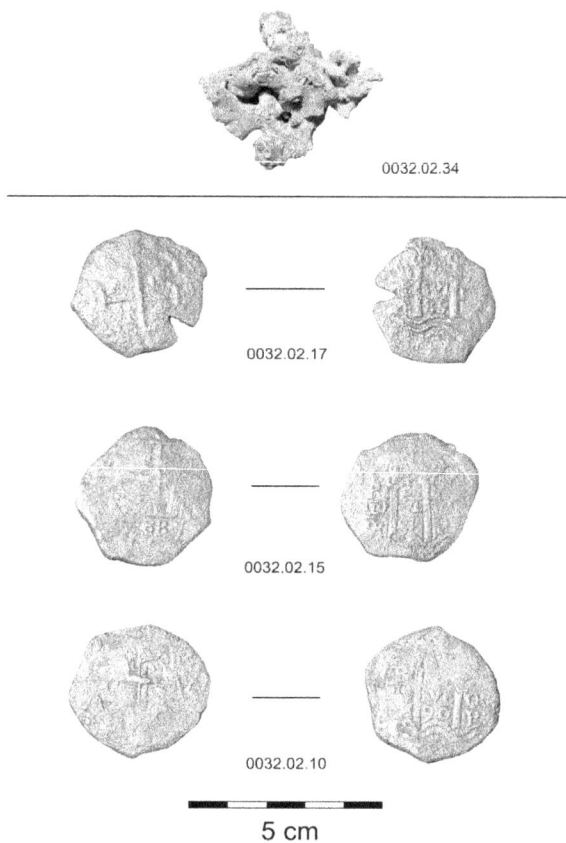

FIGURE 3. Artifacts from the Cabo Raso site: three silver coins 0032.02.10; 0032.02.15; 0032.02.17, and a fragment of red coral 0032.02.34. (Photos by the author, 2022.)

The collection from the wreck also includes 16 silver coins. Some of them are in poor condition, but it was still possible to identify them as pieces of eight (Figure 4). On the obverse, two pillars representing the Pillars of Hercules are present, with the waves below, representing the Atlantic Ocean. The letter "P", used on coins minted in Potosi, appears next to the 8 for the denomination of

"8 reale". The numeral "90" (meaning 1690) is visible between the pillars at the crest of the wave. Through the center runs the phrase PLV SVL TRA ("Plus Ultra"), meaning More Beyond the Pillars of Hercules. The Crusader's cross appears on the reverse, with lions and castles in the four quadrants representing the early Spanish kingdoms of Leon and Castile (Knecht 2019).

The various collections carried out in the 1960s provide a sample of about 55 fragments of *Corallium rubrum*, both in its natural form (raw), small branches, and pierced beads (already worked) (Figure 4). These coral fragments could have been part of the ship's cargo and could have been used by Portuguese jewelers to make various types of jewelry and/or amulets (Vieira 2020:1858).

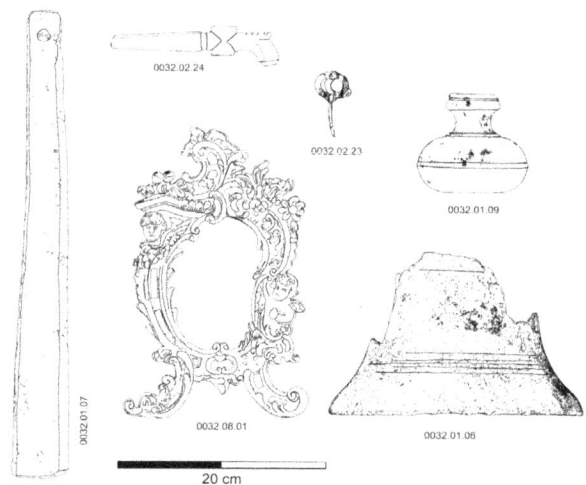

FIGURE 4. Artifacts from the Cabo Raso site: lead depth sounding 0032.01.07, brass barrel tap 0032.02.24, bronze frame 0032.08.01, cooper compass 0032.02.23, cooper candlestick 0032.01.09 and bronze bell 0032.01.06. (Drawings by the author, 2022.)

Conclusion

The Cabo Raso site is a paradigmatic case in Portuguese underwater archaeology, portraying the Portuguese State's incapacity to manage underwater cultural heritage until the 1990s. Accidently discovered, the site was the target of numerous instances of unauthorized salvage. These resulted in the disappearance of several materials, dispersed between CNANS (National Centre for Nautical and Underwater Archaeology), Museological Institutions like the Museum of the Sea in Cascais; and Museum of Angra do Heroismo (Açores), with 64% of the material. However, other finds are in private collections, which represent 36% of the collection. What happened at the Cabo Raso site is also what happened to three other wrecks: *Nossa Senhora dos Mártires*, in São Julião da Barra, Oeiras; *Slot ter Hooge*, in Porto Santo Island, Madeira wrecked on 20 November 1724; and the *L'Océan* in Salema beach in Algarve wrecked on 18 August 1759. These were discovered in the 1960s and had the same luck in their first years after the discovery. The sites were looted and subjected to the systematic recovery of material, in some cases with state permission, and, therefore, without any kind of archaeological record. This context limits the research; some finds that are in private collections have never been registered, for example.

Furthermore, although management policies for underwater cultural heritage in Portugal have improved since the 1980s (the first phase of the professionalization of underwater archaeology), some of these sites remain outside the country's capacity to act. The Cabo Raso site is once again a good example. Its location, in an area where diving is particularly difficult, has prevented systematic investigation.

However, the systematic study of the materials and documentation is an essential step in enhancing the value of this site. The finds from the Gran Principessa Di Toscana are a diverse collection of artifacts, from artillery (of Florence, Italy, and Dutch origins) to everyday life objects of the ship. They contribute to understanding the early-modern trade, namely Luso-Italian relations of the 17th century. For example, the cargo of the red coral, coming from the Mediterranean, is the first direct evidence of trade essential to the work of craftsmen that produce jewelry and religious pieces in Lisbon. After all these years, and without being properly studied, the Cabo Raso site still has an immense archaeological potential.

Acknowledgments

This paper had the support of CHAM (NOVA FCSH / UAc), through the strategic project sponsored by FCT (UIDB/04666/2020). The author would like to thank José Bettencourt for his collaboration and for reviewing the article, and Catarina Rosa for reviewing the English version.

References

ALVES, FRANCISCO J. S.
1994   *Lisboa Submersa. Lisboa Subterrânea (Underground Lisbon)*, Museu Nacional de Arqueologia, Electa, Lisboa Capital Europeia da Cultura' 94, Edição Instituto Camões, Expo'98, Lisboa: 126–132.

ATANÁSIO, M. C. MENDES
1981/1982   Canhões de bronze no Museu do Mar em Cascais, fundidos em Florença (Bronze Cannons at the Museum of the Sea in Cascais, Cast in Florence). *Arquivo de Cascais, Boletim Cultural do Município* No3: 9–16.

BETTENCOURT, JOSÉ, INÊS P. COELHO, CRISTÓVÃO FONSECA, GONÇALO LOPES, PATRÍCIA CARVALHO, AND TIAGO SILVA
2018   Entrar e sair de Lisboa na Época Moderna: uma perspectiva a partir da arqueologia marítima (Entering and Leaving Lisbon in the Modern Age: a Perspective from Maritime Archaeology). *Meios Vias e Trajetos… entrar e sair de Lisboa. Fragmentos de Arqueologia de Lisboa 2*. Câmara Municipal de Lisboa, Direção Municipal de Cultura, Departamento de Património Cultural, Centro de Arqueologia de Lisboa Sociedade de Geografia de Lisboa, Secção de Arqueologia. Lisboa: 146–161.

BORGES, MARCO OLIVEIRA
2014   Portos e Ancoradouros do Litoral de Sintra – Cascais. Da Antiguidade à Idade Moderna (I) (Ports and Anchorages of Sintra - Cascais Coast: From Antiquity to the Modern Age). *Jornadas do Mar. Mar: uma Onda de Progresso*, Almada, Escola Naval. Lisboa: 152–164.

CARDOSO, JOÃO PEDRO
2012   *Sobre os destroços da Gran Principessa Toscana, naufragada em 1696 nas imediações do Cabo Raso (On the Shipwreck of the Gran Principessa Toscana, Wrecked in 1696 near Cabo Raso)*, Cascais. Cascais, Junta de Freguesia de Cascais.

DIRECTORATE-GENERAL FOR CULTURAL HERITAGE (DGPC)
2010   MatrizNet. Direção-Geral do Património Cultural. <http://www.matriznet.dgpc.pt/>. Accessed 25 February 2023.

ESMERALDO, ANA MARIA BRANDÃO
2022   A caminho do Oriente. Os vestígios do Slot ter Hooge (1724) na dinâmica na navegação da Companhia Holandesa das Índias Orientais (Eastbound. The Remains of the Slot Ter Hooge (1724) within the Dutch East India Company Navigational Dynamics). Master's thesis, Faculdade Ciências sociais e Humanas, Universidade Nova de Lisboa. Lisboa, Portugal.

FILE 1966/01
1966   *Gran Principessa di Toscana*, Biblioteca de Arqueologia, Palácio Nacional da Ajuda, Lisboa.

FREIRE, JORGE, AND ANTÓNIO FIALHO

2013a   A Paisagem Cultural Marítima de Cascais: o
         Modelo de Investigação e de Gestão do Litoral
         (Cascais' Maritime Cultural Landscape: the Model
         for Research and Management of the Coast).
         Arqueologia em Portugal - 150 Anos. Associação dos
         Arqueólogos Portugueses, Lisboa: 1213–1220.

2013b   Cabo Raso I, Relatório ProCASC, resultados de
         2013. In *Projecto Carta Arqueológica do Concelho de
         Cascais 2013.* CHAM-CMC, Lisboa: 21–32.

FREIRE, JORGE, JOSÉ BETTENCOURT, AND ANTÓNIO FIALHO

2012   Sistema de Informação Geográfica na Gestão do
         Património Cultural Subaquático: a experiência
         da Carta Arqueológica de Cascais (Geographic
         Information System in the Management of
         Underwater Cultural Heritage: the Experience of
         the Archaeological Chart of Cascais). *2as Jornadas
         de Engenharia Hidrográfica.* 20–22 de Junho de
         2012, Lisboa.

KNECHT, ROBERT

2019   How to Read a Spanish Reale… the "Piece of 8".
         In Cannon Beach Treasure Company. <https://
         cannonbeachtreasure.com/blogs/news/how-to-
         read-a-spanish-piece-of-8-pillars-waves-edition>.
         Accessed 23 February 2023.

RIDELLA, RENATO G., EHUD GALILI, DEBORAH CVIKEL, AND
BARUCH ROSEN

2016   A Late 16th to Early 17th century European
         Shipwreck Carrying Venetian Ordnance Discovered
         off the Carmel Coast, Israel. *International Journal of
         Nautical Archaeology* 45(1):180–191.

SALGADO, AUGUSTO

2002–2006   Artilharia Naval (Naval Artillery). In
         Navegações Portuguesas. Instituto Camões. <http://
         cvc.instituto-camoes.pt/navegaport/index1.html >.
         Accessed 20 February 2023.

VIEIRA, ALEXANDRA

2020   Os Amuletos em Portugal - dos objetos às
         superstições: o coral vermelho (Amulets in
         Portugal - From Objects to Superstitions: the
         Red Coral). *Arqueologia em Portugal – Estado da
         Questão.* Associação dos Arqueólogos Portugueses.
         CITCEM. Lisboa:1849–1864.

. . . .

Sofia Simões Pereira
CHAM – Center for the Humanities
and History Department, FCSH
Universidade NOVA de Lisboa
Avenida Berna 26 C,
Lisbon, Portugal 1069-061

The Underwater Deposit of Roman Building Clay Weights of the Moaña Port (Northwest Spain)

Laura Casal Fernández, Víctor José Barbeito Pose

Roman building clay constitutes an ideal material for making weights by trimming fragments of tegula, imbrex, *and later to an appropriate size and shape, and adding perforations or lateral notches, for suspension. These types of weights have been documented in Roman coastal sites directly linked to fishing across the Atlantic-Mediterranean region. This paper presents a collection of 62 Roman building clay weights that were recovered from the coastal inlet of the Ría de Vigo (an estuary in Galicia, Spain) in 2005, and the presence of traces of synthetic rope tied around two artifacts introduces a certain amount of controversy concerning their interpretation.*

A cerâmica de construção romana constitui um material ideal para o fabrico de pesos, cortando fragmentos de tegula, imbrex, *e later, de forma e tamanho adequados, e acrescentando perfurações ou entalhes laterais para suspensão. Este tipo de pesos foi documentado em sítios costeiros romanos diretamente ligados à pesca em toda a região atlântico-mediterrânica. Este artigo apresenta uma coleção de 62 pesos pesos em cerâmica de construção romana que foram recuperados da enseada costeira da Ría de Vigo (um estuário na Galiza, Espanha) em 2005. A presença de vestígios de corda sintética atada em torno de dois artefactos introduz uma certa controvérsia quanto à sua interpretação.*

L'argile de construction romaine constitue un matériau idéal pour fabriquer des poids en découpant des fragments de tegula, *d'*imbrex, *et de later à une taille et une forme appropriées, et en ajoutant des perforations ou des encoches latérales, pour la suspension. Ces types de poids ont été documentés dans des sites côtiers romains directement liés à la pêche dans la région Atlantique-Méditerranée. Cet article présente une collection de 62 poids d'argile de construction romaine qui ont été récupérés dans l'anse côtière de la Ría de Vigo (un estuaire en Galice, en Espagne) en 2005, et la présence de traces de corde synthétique attachée autour de deux artefacts introduit une certaine controverse concernant leur interprétation.*

Introduction: Geographical Location and Roman Archaeological Context

The Port of Moaña (Galicia, Spain) is located on the inlet of the same name, on the northern shore of the Ría de Vigo, which lies on the coastal stretch known as the Rías Baixas in Galicia. This region in the northwest section of the Iberian Peninsula's Atlantic coast stands out for the wealth of its fishing and shellfish resources, a fact which was already well-known in Roman times, judging by the testimony offered in the writings of the classical author, Martial (*Epigramas* X 37). The original layout of the maritime and terrestrial space occupied by the Port of Moaña has been profoundly altered by a series of works carried out on the port (landfill to the sea, breakwater, and harbor works) between the first half of the 20th century and the present (Figure 1). The long and deep-rooted tradition of exploiting the marine resources of the Ría de Vigo reaches back to antiquity. The archaeological record of numerous coastal hill forts

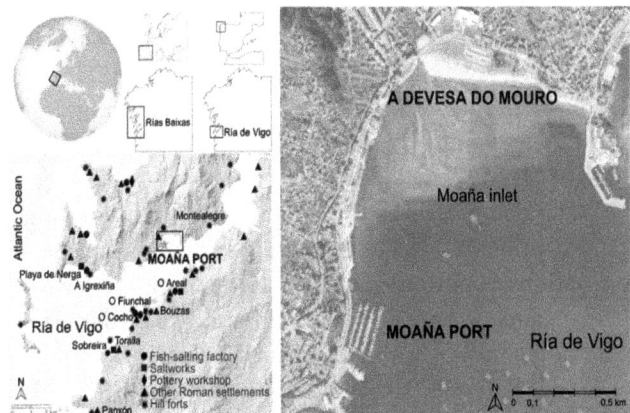

FIGURE 1. Location of Moaña Port in Northwest Spain. Left: Map of Roman occupation of the Ría de Vigo coast with sites mentioned in the text. Right: Details of the dredged area in Moaña Port and of the close Roman site of A Devesa do Mouro (PXOM Moaña 2016). Cartographic bases and archaeological sites data made available by Información Xeográfica de Galicia (Plan Básico Autonómico (Galicia) [PBA] 2022). North arrows represent true north. (Figure created by the primary author, 2023.)

indicates that the reliance on marine resources for sustenance in pre-Roman times was relatively high.

This is shown by the zooarchaeological remains (ichthyofauna and malacofauna), which are sometimes concentrated in large shell middens (known as *kjokkenmoedding* in the archaeological historiography of the late 19th and early 20th centuries), as well as by fishing instruments (González 2006–2007:298–302; González 2013; Bejega 2015). Following the Roman conquest of the northwestern Iberian Peninsula, the Rías Baixas coastal region experienced a growing, economically driven occupation aimed at exploiting the marine environment. This process was favored by particularly suitable physical conditions for human settlement and navigation, with sandy and marshy areas, sheltered bays, and natural harbors. The archaeological evidence from the Ría de Vigo clearly supports this: four saltworks (Punta de Toralla, Bouzas, O Areal, and Playa de Nerga/As Forcadas creek) (Castro 2007; Pérez et al. 2008:195–198, 502–503; Gorgoso and Acuña 2016:73, 93, figure 3; Currás 2017; Casal 2018:300–303); five fish-salting factories (Sobreira, O Cocho/Punta Borralleiro, O Fiunchal, O Areal, and A Igrexiña) (Hidalgo 1990–1991:195–196; Castro 1992–1993; Hidalgo and Rodríguez 1995; Torres et al. 2007; López 2009, 2010; Gorgoso and Acuña 2016; Currás 2017; Fernández 2017, 2018); and two *villae a mare* (Panxón and Toralla) (Pérez et al. 2008; Pérez 2009; Villar and Villacieros-Robineau 2010; Acuña 2013:146). Other coastal settlements of Roman origin and whose exact nature cannot always be ascertained, have been discovered (Pérez et al. 2008: figure 1), including some that show evidence of fishing and shellfish harvesting (Villar 2008; Casal 2022:198–201, 338–341).

The typology of weights obtained through trimming and adapting miscellaneous pottery fragments (amphorae, common pottery, and Roman building clay) with perforations or lateral notches is included in the classification of fishing instruments used in antiquity as proposed by Bernal (2010:101–103, figure 1). Their presence has been recorded in different coastal regions of the Iberian Peninsula displaying thriving fishing and related activities during antiquity, such as the Circle of the Strait (Vargas 2020:71–73); the area surrounding the ancient city of Olisipo and the large fish-salting plants of Tróia on the Portuguese Atlantic seaboard (Mayet and da Silva 2000:66, figures 28–167); or ancient Barcino in the northeast of the Iberian Peninsula (Beltrán 2007:279, figures 4-3, 5). Depending on their size, weight, and perforation diameter, trimmed clay weights have been regarded as depth-adjusting devices for various fishing gear, such as single- and multiple-hooked lines, as well as nets and creels, either simple or line linked (Bernal 2010:102; Vargas 2020:73).

A significant amount of Roman building clay weights used for suspensory purposes and made from *tegula*, or Roman flat tile; *imbrex*, or Roman curved tile; and *later*, or Roman brick, have been documented in the Ría de Vigo area. These items were found on coastal sites of Indigenous origins, such as the hill fort of Montealegre, where 23 specimens were discovered, as well as settlements of clearly Roman origin, such as the *villa* of Toralla or O Areal, yielding 6 and 56 items, respectively (Casal 2018:304–305; 2022:206, 238, 247–248, 500–502, figures 173–175). However, attributing these types of clay artifacts to fishing is not without its problems in the case of items recovered on land, as they might also have been used as loom weights in the weaving industry. In any case, it should be pointed out that while the heterogeneous shapes and sizes, as well as the strongly worn surfaces of certain items, do not guarantee their functional attribution to fishing, they nevertheless provide strong support for this hypothesis.

2005 Archaeological Control Project: Description and Results

In 2005, the dredging works carried out on the seabed of Moaña Port to accommodate the planned sports facilities were accompanied by an archaeological control and monitoring project (Barbeito 2005). The dredging affected a rectangular area of 22,843.80 square meters (m²), where a total of 77 extractions were performed. The methodology applied during this preventive archaeological measure was mainly limited to a continuous visual inspection of the dredged materials.

The area that was examined corresponds to a sedimentary marine environment, with a clearly aggrading sequence. Based on the granulometric properties of the seabed surface, three phases are clearly distinguishable: the area closest to the coast, which coincides with the highest level, contains a high percentage of coarse sand and gravel, containing shell remains (SP01); the next phase corresponds to medium- and fine-fraction sands associated with more fragmented mollusk shell remains (SP02); finally, the third phase is characterized by the predominance of terrigenous materials (mainly silts and mud), and an abundance of finely crushed shells (SP03). In general, the materials originating from the SP01 zone are less eroded than those from the areas furthest from the coastline.

Dredging alters the structure of the seabed as it deforms the preexisting stratification. Given the impossibility of contextualizing the collected remains of material culture from a stratigraphic point of view, a criterion based on areas of concentration and dispersion of materials was applied (Figure 2). The area showing the highest density corresponds to the section closest to the coastline (SP01), where the presence of the current port structures favors access and movement, thus converting it into a recipient area for waste and lost materials. More specifically, the highest concentration was identified in the area that matches the current port and adjacent reclaimed land, while it gradually decreases towards the north and the east. The areas showing higher concentrations tend to lie to the southwest due to predominating currents, the latter encountering physical barriers in the port structures (reclaimed land, breakwater, and harbor) preventing normal movement.

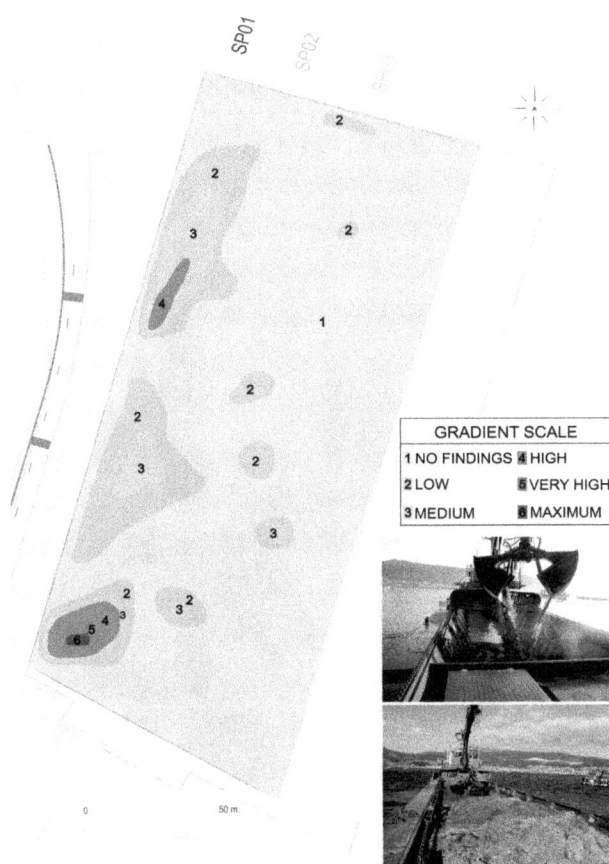

FIGURE 2. Plane of correlation between density graph and sedimentary sequences. Concentration areas of materials and their orientation. North arrow represents true north. Images of dredging works. (Figure from Barbeito 2005:figure 5, photographic documentation.)

The available data suggests a continuous resedimentation process within a low-energy environment that finds itself locked in by the port structures. This directly results in clustering processes whereby various sources (isolated events that occur when material culture items are lost) contribute materials that gravitate towards a common point where it is deposited in growing concentrations. The nature of the archaeological record supports the hypothesis of clustering processes, as it comprises a heterogeneous set of materials, originating from different historical periods, displaced and mixed within the same archaeological entity. The accumulation of this artifact assemblage implies a human input of waste and lost materials that is later altered by natural and environmental variables (e.g., the topography of the seabed or marine currents), as well as by human and cultural ones (e.g., building of port structures or shellfish-harvesting). The result is the formation of a new entity or formative context (Renfrew and Bahn 1993:43–63), which is purely natural, rather than cultural, i.e., devoid of any cultural stratigraphic context. The remains of material culture that were recorded and studied show that the area was the scene of, more or less, uninterrupted human activity of varying intensity during a certain period, as indeed it continues to be.

Material Culture: Roman Building Clay Weights

The 77 extractions carried out on the area under archaeological scrutiny in the Port of Moaña resulted in the recovery of an assemblage of 218 items belonging to different historical periods. This did not include any building remains of note. Most of the archaeological record consists of hand-crafted or industrial clay fragments of relatively recent manufacture (19th and 20th centuries). Occasional items dating back to a period ranging between the 16th and 18th centuries were also noted. A significant part of the assemblage—nearly a third—consists of clay fragments from the Roman to Late Roman period, the majority of which (91.2%) are Roman building clay fragments trimmed and suitably adapted to a new purpose by applying perforations or lateral notches allowing them to be tied to a line. In addition, two pebbles with lateral notches were identified that were undoubtedly used as weights for fishing tackle or gear (Figure 3f). These types of stone weights have been documented across the northwestern Iberian Peninsula from the Late Bronze Age/Early Iron Age to the 20th century, while their massive use as fishing weights has been attested since the Early Modern period

(Rodríguez 1923:578, 674–676; Vázquez 2000:53–55; Casal 2022:81–85-, 511–516).

The assemblage of Roman building clay found on the seabed of Moaña Port is made up of a total of 62 fragments. Two main types of weights can be distinguished according to the system used for suspending or tying: perforated items with a single hole (32.3%) (Figure 3a) and weights with lateral notches (35.5%) (Figure 3b). A significant part of the assemblage (29%) is made up of items without perforations or clearly visible notches, which raises initial doubts as to their use for suspensory purposes. However, the presence of possible tying traces on the surface of certain specimens, the fact they were discovered close to one another, and similar specimens with an unmistakable suspensory function account for the decision to include them in the current study. The remaining 3.2% are represented by two *tegula* fragments, with traces of unfinished perforations, which have been interpreted as possible pre-shaped or unfinished examples of perforated trimmed weights.

Perforated Weights

Regarding the assemblage of perforated weights, this discussion looks at various factors, such as the weight, size, hole diameter, shape, and the type of fragment of Roman building clay used for their manufacture (Table 1). The total of 20 items constituting the set of perforated weights presents a wide range of weights between 160.17

and 799.50 grams (g), with the items' thickness rather than their length accounting for the weight, which is ultimately determined by the type of Roman building clay fragment selected for making the artifact. The thickness varies considerably between items obtained from *imbrex*, or from the flat part of *tegula* (2.10 centimeters [cm] or more), and those made from fragments of *later*, or from *tegula* including the raised edge (up to 5.05 cm). Thus, the range of weights of the artifacts stands in contrast to the limited range of lengths, which ranges between 8.75 and 13.15 cm. Regarding the size of the hole serving to secure the weight to the fishing tackle or gear, the set of weights recorded and studied includes diameters ranging between 0.4 and 1.35 cm.

Likewise, the set of perforated weights displays a wide range of shapes, leading to the following classification in decreasing order of frequency: near-oval (35%), near-triangular (25%), near-trapezoidal (15%), near-rectangular or near-quadrangular (15%), near-rhomboid (5%), and near-pentagonal (5%). A clear predominance of weights is made from *tegula* fragments (70%), whereas only a minority of items is made from fragments of *later* (15%) and *imbrex* (5%). The other group of weights, 10% of the total, is problematic in terms of determining the kind of recycled Roman building clay employed, *later* or raised edge sections of *tegula*, both having a similar thickness.

FIGURE 3. Roman building clay: (a) perforated weights; (b) notched weights; (c) *tegula* with remains of unfinished perforations; (d) *tegula* with notches on the edge and detail of tying trace; (e) possible weight obtained from a *tegula* fragment. (f) pebble with lateral notches. (Photos and drawings by the primary author, 2021.)

Shape	No.	Roman Building Clay Fragment	Weight (g) min.-max.	Hole diameter (cm) min.-max.	Length (cm) min.-max.
Near-oval	7	4 *tegula*, 1 *imbrex*, 1 *later*, 1 indeterm.	163.9–413.8	0.6–1.35	10.9–>13.15
Near-triangular	5	4 *tegula*, 1 *later*	198–298.3	0.55–1.05	10.15–11.9
Near-trapezoidal	3	3 *tegula*	160.1–392	0.4–1.2	9.3–12.1
Near-rectangular or near-quadrangular	3	1 *tegula*, 1 *later*, 1 indeterminate	262–799.5	0.9–1.25	8.75–17.3
Near-rhomboid	1	1 *tegula*	205.07	0.65	12.5
Near-pentagonal	1	1 *tegula*	209.98	0.55	10.65
Total	20	14 *tegula*, 1 *imbrex*, 3 *later*, 2 indeterm.	-	-	-

TABLE 1: Inventory of Roman building clay perforated weights.

Shape	No.	Roman Building Clay Fragment	Weight (g) min.-max.	Length (cm) min.-max.
Near-rectangular or near-quadrangular	15	13 *tegula*, 2 *imbrex*	116.74–841.6	5.6–17.9
Irregular	4	4 *tegula*	97.36–399.53	5.0–14.9
Near-oval	1	1 *tegula*	202.48	12.35
Near-pentagonal	1	1 *tegula*	276.35	9.9
Star-shaped	1	1 *tegula*	136.02	8.7
Total	22	20 *tegula*, 2 *imbrex*	-	-

TABLE 2: Inventory of Roman building clay weights, with notches.

The relative evenness of size documented for perforated trimmed artifacts could be due to a manufacturing method based on relatively rudimentary techniques giving rise to weights of variable shapes. The exceptional finding of two sizable *tegula* fragments, with traces of unfinished perforations (Figure 3c), provides hints as to the process employed in obtaining perforated trimmed weights from Roman building clay material. Curiously, the best-preserved *tegula* presents two cavities on the flat section and a third on the transition between the raised edge and the flat section. These positions match the perforations found on the perforated weights recorded.

It should be pointed out that during the fieldwork two perforated weights were identified that were tied to a piece of synthetically made rope by means of a thin piece of string holding them together through the perforations. The presence of synthetic thread complicates the interpretation of the assemblage, as it introduces a *terminus post quem* in the second third of the 20th century.

Notched Weights

The set of trimmed weights, with lateral notches, is represented by a total of 22 specimens. The following aspects were examined: weight, size, shape, and type of repurposed Roman building clay fragment used (Table 2). The weight range of the notched fishing weights is slightly wider than that of the perforated weights, varying between 97.36 and 841.6 g. Unlike the perforated weights, the notched weights show a proportional relationship between their weight and length. In terms of shape, a certain degree of uniformity is present in that near-rectangular and near-quadrangular items clearly predominate (68.2%). The rest includes irregularly shaped (18.2%), near-oval (4.5%), and near-pentagonal weights (4.5%), in addition to one star-shaped weight (4.5%), which illustrates better than any other item the crosswise tying pattern that can be inferred from the position of the notches found on the majority of weights. On the other hand, the vast majority of notched weights seem to be made from repurposed *tegula* fragments (91%), as compared to the only two specimens made from *imbrex* (9%).

It should be noted that rope traces are present on the surface of certain notched weights (Figure 3d). Examining these traces, together with the position of the notches along the edges, presents hypotheses regarding the type of system employed for tying and how the weights would have been secured to the fishing tackle or gear (Figure 4). Some of the notches on the weight edges seem to have been produced by wear, resulting from the rope tied around them. The erosion produced by the water, in combination with the sand of the seabed, would have progressively worn away any sharp points and projections giving rise to their rounded aspect.

As mentioned above, 29% of Roman building clay fragments documented on the seabed of Moaña Port corresponds to items with an uncertain function (Figure 3e). This group comprises a total of 18 items characterized by the presence of slight depressions on the edge pointing to a possible suspensory function and, thus, supporting the hypothesis of their use as fishing weights, like their perforated and laterally notched counterparts.

Discussion and Conclusion

The discovery of 62 Roman building clay fragments on the seabed of the Galician Port of Moaña, 42 of which are suitably adapted for suspending by means of perforations or lateral notches, leaves no doubt as to the repurposing of this set of clay materials for use as fishing weights. Unlike artifacts found on land sites, the underwater context of the find guarantees its attribution to fishing activity. However, the presence of remnants of synthetic rope found attached to two perforated items is disconcerting and poses a problem in terms of interpreting the assemblage as a whole. This unexpected finding confirms the 20th-century reutilization of Roman or Late Roman clay materials in a geographic environment, with a high rate of occupation during antiquity where the strong tradition of exploiting marine resources has continued to the present day. Given the fragility of the

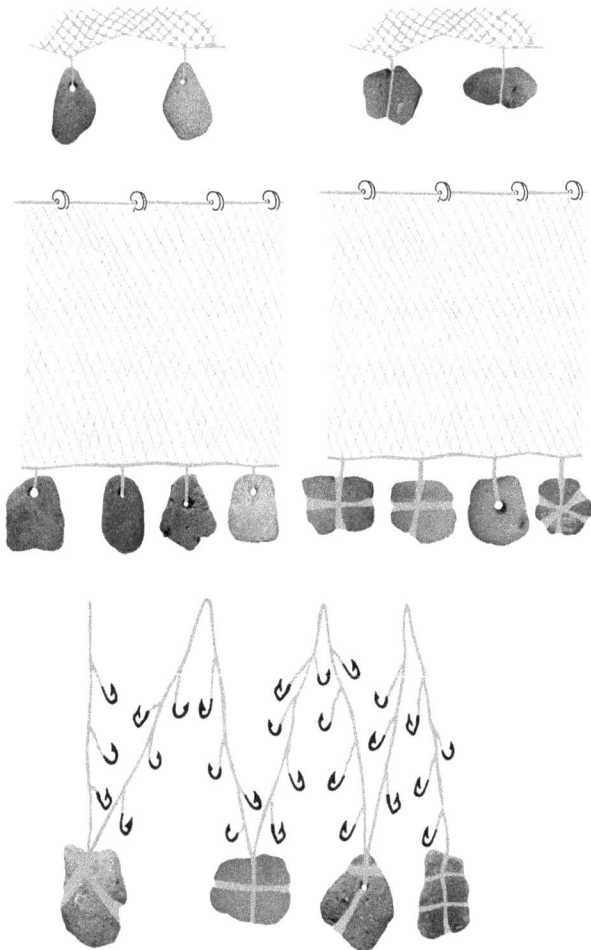

FIGURE 4. Hypothetical examples of nets, fishing tackle, and systems used for tying. (Figure created by the primary author, 2023, and inspired by Lorenzo 1982:59, 149.)

primary material, it can be ruled out that Roman building clay weights survived from antiquity until the 20th century. Therefore, one may infer that at some point during the second third of the 20th century, the local traditional fishing community provided itself with pre-existing clay items to manufacture their own fishing gear and tackle. In fact, recycling of contemporary building clay fragments as fishing weights is well documented in Spain during the Early and Late Modern periods (Sáñez 1793:342). Indeed, this kind of material is ideally suited for obtaining weights, as it is easy to work with rudimentary tools, in addition to its easy accessibility and affordability, and the speed of replacement in case of loss or breakage. More specifically, Roman building clay enables fishers to obtain weights of more or less appropriate shapes and sizes for each individual application, as it can be adapted to specific needs, which fits with the highly pragmatic and opportunistic nature of this craft-based activity.

Building on the assumption from the previous paragraph, namely that local fishers started reusing Roman building clay fragments to make their fishing gear sometime after the second third of the last century, the question arises as to the source of the clay material. The most plausible hypothesis is that materials encountered by accident at the Roman site of A Devesa do Mouro, located in the current town center of Moaña next to the port, were deliberately collected, notwithstanding alternative sourcing from other coastal sites in the vicinity. This hypothesis is strengthened by the record of two *tegulae*, with traces of several incomplete perforations, which have been interpreted as weights in the process of being manufactured or left unfinished. In reality, what is known of the site of A Devesa do Mouro derives from an area of dispersion where remains of clay material—essentially *tegulae* and amphorae—were discovered during various home development works and farming-related activities, with the consequent altering of the archaeological site (Plan Xeral de Ordenación Municipal [PXOM] Moaña 2016).

Comparing the results obtained from archaeological control projects linked to recent dredging operations in some of the traditional Galician ports, such as the inlet of Baiona (Pontevedra) (San Claudio 2009), the bay of A Coruña (San Claudio 2007), or the Port of Ares (A Coruña) (San Claudio 2004), a conspicuous difference with Moaña Port exists, namely the marked predominance of stone weights—mainly pebbles, with lateral notches (widely used in Galicia since the Early Modern period)—against a minority or absence of clay weights in the case of the other locations mentioned. It may be assumed that the substantial presence of Roman building clay weights in Moaña Port is due to a one-off and circumstantial event.

While the archaeological data available is not sufficient to extrapolate the discovery of synthetic rope remains found on two items to the whole set of weights recorded at Moaña Port, this supposition cannot be completely dismissed. Indeed, it could be argued that it offers sufficient ground for questioning the assumption that attributes the scarce number of Roman building clay weights retrieved from underwater archaeological works in Galicia to the Roman period, both in the Rías Baixas (Peña Santos 1985:214–215; Casal 2022:298) and on Galician coastal stretches located further north (López 1980:150, 155, figure 18).

Few archaeological parallels exist to draw on when it comes to Roman building clay weights recovered from underwater contexts, and hardly any archaeological historiography is available covering this line of research—fishing archaeology and, more specifically, the study of fishing implements—as it is still a young discipline. Even Atlantic-Mediterranean regions that have been the subject of studies to compile a *corpus* of ancient fishing equipment, such as the Circle of the Strait and the East Coast of the Iberian Peninsula, have not yielded any Roman building clay weights from underwater contexts, only from terrestrial archaeological sites (Vargas 2020:71–73). One of the closest parallels is the assemblage of 91 clay weights without a defined context (shipwreck, submerged site, or port structures) that were retrieved from the sea in the vicinity of the large fish-salting plants of Tróia, near the mouth of the Sado River, and in Quarteira, on the Portuguese Algarve, and that have been interpreted as fishing net weights (Loureiro and Martinho 2003). The authors used a type-chronological classification, which distinguishes six categories of weights whose dating extends from the Roman period to the 20th century. However, weights made of reused Roman building clay do not feature among the 91 recovered pieces.

In short, the archaeological record of Moaña Port constitutes an unprecedented discovery. Irrespective of the chronological attribution, the unexpected underwater discovery in Moaña Port proves the feasibility of reusing Roman building clay as fishing weights. In truth, it is difficult to argue against an extrapolation of this phenomenon to the Roman period, especially considering the archaeological evidence concerning the use of clay weights in ancient nets and fishing gear (Bernal 2010:98–103, figure 1).

It is not known what type of fishing gear these weights would have been tied to, but the finding of two perforated items tied to one main rope (in this case made of synthetic fiber) by means of strands that were threaded through the perforations suggests that they were used together with nets or with multiple-line rigs, consisting of a main line branching into several secondary lines holding hooks, creels or a combination of both. The overall diversity of the pieces may point either to the use of different-sized nets and rigs or to a poor or hasty workmanship, lacking in regularity.

The find of various specimens showing clear rope-wear signs on their surface invites future research focused on traceology. Use-wear marks analysis could shed some additional light on the composition of the tying ropes and on the fastening system to the nets and fishing tackle. Perhaps future underwater archaeology works in the Ría de Vigo or other areas showing evidence of marine resource exploitation in antiquity, combined with greater research efforts on the study of fishing instruments, will allow this recent and, as yet, scarcely studied research line to move forward.

References

Acuña Castroviejo, Fernando
2013   De novo sobre o mosaico de Panxón e outras novas sobre a musivaria na Gallaecia (Again about the Mosaic of Panxón and Other News about Mosaics in Gallaecia). *Revista da Faculdade de Letras, Ciências e Técnicas do Património* XII:143–-157 <https://ler.letras.up.pt/uploads/ficheiros/11794.pdf>. Accessed 21 December 2022.

Barbeito Pose, Víctor
2005   Control arqueolóxico das obras de dragaxe de instalacións para navegación recreativa en Moaña (Moaña, Pontevedra). Informe-Memoria técnica (Archaeological Control of the Dredging Works for Sport Navigation Facilities in Moaña [Moaña, Pontevedra]). Technical Report. Report to Dirección Xeral de Patrimonio Cultural, Xunta de Galicia, Santiago de Compostela, España, from Adro Arqueolóxica S.L., Santiago de Compostela, ES.

BEJEGA GARCÍA, VÍCTOR
2015   El marisqueo en el Noroeste de la península Ibérica durante la Edad del Hierro y la época romana (Shellfishing in the Northwest of the Iberian Peninsula during the Iron Age and Roman Times). Doctoral dissertation, Departamento de Historia, Universidad de León, España <https://buleria.unileon.es/handle/10612/5126>. Accessed 21 December 2022.

BELTRÁN DE HEREDIA, J.
2007   *Cetariae* bajo imperiales en la costa catalana: el caso de *Barcino* (Late Empire *Cetariae* on the Catalan coast: the Case of *Barcino*). In *Salsas y salazones de pescado en Occidente durante la Antigüedad, Actas del Congreso Internacional (Cádiz, 7-9 de noviembre de 2005)*, L. Lagóstena, D. Bernal, and A. Arévalo, editors, pp. 277–284. British Archaeological Reports International Series 1686, Oxford, United Kingdom <https://www.academia.edu/29446737/_Cetariae_bajo_imperiales_en_la_costa_catalana_el_caso_de_Barcino_2007>. Accessed 21 December 2022.

BERNAL CASASOLA, DARÍO
2010   Fishing Tackle in Hispania: Reflections, Proposals and First Results. In *Ancient Nets and Fishing Gear. Proceedings of the International Workshop on "Nets and Fishing Gear in Classical Antiquity: A First Approach"*, Tønnes Bekker-Nielsen and Darío Bernal Casasola, editors, pp. 83–137. Cádiz, ES.

CASAL FERNÁNDEZ, LAURA
2018   La *villa* romana de Toralla y la explotación de los recursos marinos en las Rías Baixas (The Roman Villa of Toralla and the Exploitation of Marine Resources in the Rías Baixas). *Actas de las IX Jornadas de Jóvenes en Investigación Arqueológica, Santander 8-11 junio 2016*, Lucía Agudo Pérez, Carlos Duarte, Asier García Escárzaga, Jeanne Marie Geiling, Antonio Higuero Pliego, Sara Núñez de la Fuentes, Fco. Javier Rodríguez Santos, and Roberto Suárez Revilla, editors, pp. 299–306. Instituto Internacional de Investigaciones Prehistóricas de Cantabria, Santander, España <https://www.researchgate.net/publication/335389903_Actas_de_las_IX_Jornadas_de_Jovenes_en_Investigacion_Arqueologica>. Accessed 21 December 2022.

2022   La explotación de los recursos marinos en época romana en el Noroeste de la península Ibérica (The Exploitation of Marine Resources in Roman Times in the Northwest of the Iberian Peninsula). Doctoral dissertation, Departamento de Historia, Arte y Geografía, Universidad de Vigo, España <https://www.investigo.biblioteca.uvigo.es/xmlui/handle/11093/3031>. Accessed 21 December 2022.

CASTRO CARRERA, JUAN C.
1992–1993   Intervención arqueolóxica no xacemento de "O Fiunchal" (Alcabre, Vigo) (Archaeological Intervention at the Site of "O Fiunchal" [Alcabre, Vigo]). *Castrelos* 5–6:71–86.

2007   La salina romana del yacimiento de "O Areal", Vigo (Galicia): un complejo industrial salazonero altoimperial (The Roman Saltwork of "O Areal", Vigo (Galicia): an Early Empire Fish-salting Complex). In *Salsas y salazones de pescado en Occidente durante la Antigüedad, Actas del Congreso Internacional (Cádiz, 7-9 de noviembre de 2005)*, L. Lagóstena, D. Bernal, and A. Arévalo, editors, pp. 355-365. British Archaeological Reports International Series 1686, Oxford, United Kingdom <https://www.academia.edu/4892796/La_salina_romana_del_yacimiento_de_O_Areal_Vigo_Galicia_un_complejo_industrial_salazonero_altoimperial>. Accessed 21 December 2022.

CURRÁS REFOJOS, BRAIS
2017   The *Salinae* of O Areal (Vigo) and Roman Salt Production in NW Iberia. *Journal of Roman Archaeology* 30:325–349 <https://digital.csic.es/bitstream/10261/230546/4/The_salinae_of_O_Areal.pdf>. Accessed 21 December 2022.

FERNÁNDEZ FERNÁNDEZ, ADOLFO
2017   Praia de Sobreira (Oia) (Vigo, España). Red de excelencia atlántico-mediterránea del patrimonio pesquero de la Antigüedad <http://ramppa.uca.es/cetaria/praia-de-sobreira-oia>. Accessed 21 December 2022.

2018   Intervención arqueológica de urgencia en la factoría de salazón romana de Sobreira (Vigo). Memoria interpretativa (Emergency Archaeological Intervention in the Roman Fish-salting Factory of Sobreira (Vigo). Interpretive Report). Report to Servizo de Arqueoloxía do Concello de Vigo, España, from Grupo de Estudios de Arqueología, Antigüedad y Territorio, Universidad de Vigo, ES.

GONZÁLEZ GÓMEZ DE AGÜERO, EDUARDO
2013   La ictiofauna de los yacimientos arqueológicos del Noroeste de la península Ibérica (Fishes in Archaeological Sites in the Northwest of the Iberian Peninsula). Doctoral dissertation, Departamento de Historia, Universidad de León, España <https://buleria.unileon.es/handle/10612/3378>. Accessed 21 December 2022.

GONZÁLEZ RUIBAL, ALFREDO
2006–2007 *Galaicos. Poder y comunidad en el Noroeste de la península Ibérica (1200 a.C.-50 d.C.) (Galaicos. Power and Community in the Northwest of the Iberian Peninsula)*. Brigantium 18–19, Museo Arqueolóxico e Histórico Castelo de San Antón. A Coruña, ES.

GORGOSO LÓPEZ, LINO, AND A. ACUÑA PIÑEIRO
2016   Igrexiña fronte ao mar. Unha salgadura romana en Nerga (Cangas, Pontevedra) (Igrexiña face to the sea: a Roman fish-salting factory in Nerga [Cangas, Pontevedra]). *Gallaecia* 35:71–98 <https://revistas.usc.gal/index.php/gallaecia/article/view/4073>. Accessed 21 December 2022.

HIDALGO CUÑARRO, JOSÉ MANUEL
1990–1991 Últimas excavaciones arqueológicas de urgencia en Vigo: castros y yacimientos romanos (Latest Emergency Archaeological Excavations in Vigo: Hill Forts and Roman Sites). *Castrelos* 3–4:191–215.

HIDALGO CUÑARRO, JOSÉ MANUEL, AND EUGENIO RODRÍGUEZ PUENTES
1995   Excavación arqueolóxica de urxencia na praia do Cocho, Alcabre (Vigo, Pontevedra) (Emergency Archaeological Excavation at Praia do Cocho, Alcabre [Vigo, Pontevedra]). *Arqueoloxía Informes* 3 (Campaña 1989):165–168.

LÓPEZ, FELIPE-SENEN
1980   Arqueoloxía submariña: os materiais procedentes da badía coruñesa (Underwater Archaeology: The Materials from the Bay of a Coruña). *Brigantium* 1:139–165.

LÓPEZ RODRÍGUEZ, ENRIQUETA
2009   Sondaxes arqueolóxicas manuais na parte oriental do inmoble nº 2–3 da praza de Compostela, Vigo (ManualArchaeological Surveys in the Eastern Part of Building nº 2–3 in Praza de Compostela, Vigo). *Actuacións Arqueolóxicas* Ano 2007:176–177 <https://issuu.com/davidabella/docs/actuacions_arqueoloxicas_2007/1>. Accessed 21 December 2022.

2010   Escavación arqueolóxica en área no sector oriental do soar nº 2–3 da praza de Compostela, Vigo (Archaeological Excavation in the Eastern Sector of Plot nº 2--3 in Praza de Compostela, Vigo). *Actuacións Arqueolóxicas* Ano 2008:193–195 <https://issuu.com/davidabella/docs/actuacion_s_arqueoloxicas_2008>. Accessed 21 December 2022.

LORENZO FERNÁNDEZ, XAQUÍN
1982   *O Mar e os Ríos (The Sea and the Rivers)*. Galaxia, Vigo, ES.

LOUREIRO, VANESSA, AND CARLA MARTINHO
2003   Colecçao de arqueologia subaquática Mestre Soares Branco: pesos de rede de Tróia e Quarteira (Mestre Soares Branco Underwater Archeology Collection: Net Weights of Tróia and Quarteira). *Boletim Cultural Câmara Municipal de Mafra* 2002:244–260.

MARTIAL
1997   *Epigramas*, Juan Fernández Valverde and Antonio Ramírez de Verger, translators. Gredos, Madrid, ES.

MAYET, FRANÇOISE, AND CARLOS TAVARES DA SILVA
2000   *Le site phénicien d'Abul (Portugal). Comptoir et sanctuaire (The Phoenician Site of Abul (Portugal): Trasing Post and Sancturary)*. De Boccard, Paris, FR.

PEÑA SANTOS, ANTONIO DE LA
1985   Primeras prospecciones arqueológicas subacuáticas en el litoral de la provincia de Pontevedra (First Underwater Archaeological Surveys on the Coast of Pontevedra Province). *Pontevedra Arqueológica* 1:205–238 <http://www.grupogarciaalen.es/publicaciones/revistas/pontevedra-arqueologica-i/>. Accessed 23 December 2022.

PÉREZ LOSADA, FERMÍN E.
2009   Escavación arqueolóxica en área na vila romana de Toralla, Vigo (Pontevedra) (Archaeological Excavation in the Roman Villa of Toralla, Vigo [Pontevedra]). *Actuacións Arqueolóxicas* Ano 2007:84–85 <https://issuu.com/martinxvm/docs/informe_2007toralla>. Accessed 21 December 2022.

PÉREZ LOSADA, FERMÍN, ADOLFO FERNÁNDEZ FERNÁNDEZ, AND SANTIAGO VIEITO COVELA
2008   Toralla y las villas marítimas de la Gallaecia atlántica. Emplazamiento, arquitectura y función (Toralla and the Villae Maritimae of Atlantic Gallaecia. Location, Architecture, and Function). In *Las villae tadorromanas en el Occidente del Imperio: arquitectura y función. IV Coloquio Internacional de Arqueología en Gijón,* Carmen Fernández Ochoa, Virginia García-Entero, and Fernando Gil Sendino, editors, pp. 481–506. Gijón, España <https://www.academia.edu/18689899/053_P%C3%A9rez_Losada_F_Fern%C3%A1ndez_Fern%C3%A1ndez_A_Vieito_Covela_S_2008_Toralla_y_las_villas_mar%C3%ADtimas_de_la_Gallaecia_atl%C3%A1ntica_Emplazamiento_arquitectura_y_funci%C3%B3n_in_Fern%C3%A1ndez_Ochoa_et_al_Eds_Las_villae_tardorromanas_en_el_occidente_del_Imperio_Gij%C3%B3n_p_481_506>. Accessed 21 December 2022.

PLAN BÁSICO AUTONÓMICO (GALICIA) (PBA)
2022   Información xeográfica de Galicia. Plan Básico Autonómico. Visor. Afeccións Patrimonio Cultural <http://mapas.xunta.gal/visores/pba/>. Accessed 3 January 2023.

PLAN XERAL DE ORDENACIÓN MUNICIPAL (MOAÑA) (PXOM MOAÑA)
2016   Plan Xeral de Ordenación Municipal do Concello de Moaña (Pontevedra). Volumen 7.A. Catálogo do Patrimonio Cultural. Fichas dos elementos arqueolóxicos. Ficha nº 15 (GA 36029015) <https://siotuga.xunta.gal/siotuga/documentos/urbanismo/MOANA/documents/27581ca007.PDF>. Accessed 3 January 2023.

RENFREW, COLIN, AND PAUL G. BAHN
1993   *Arqueología. Teorías, métodos y práctica (Archaeology. Theories, Methods, and Practice) (1998).* María Jesús Mosquera Rial, translator. Akal, Madrid, ES.

RODRÍGUEZ SANTAMARÍA, BENIGNO
1923   *Diccionario de artes de pesca de España y sus posesiones. Sucesores de Rivadeneyra (Dictionary of Fishing Gear in Spain and its Possessions).* Sucesores de Rivadeneyra, Madrid, ES.

SAN CLAUDIO SANTA CRUZ, MIGUEL
2004   Memoria técnica de la prospección arqueológica terrestre subacuática para la ampliación de las instalaciones náuticas deportivas en Ares, La Coruña (Technical Report of Underwater-Terrestrial Archaeological Prospection for the Expansion of the Nautical Sports Facilities in Ares, La Coruña). Report to Dirección Xeral de Patrimonio Cultural, Xunta de Galicia, Santiago de Compostela, España, from Arqueología Marítima, A Coruña, ES.

2007   Memoria técnica de la prospección arqueológica subacuática. Instalaciones para la náutica deportiva en San Antón, La Coruña, 2006–2007 (Technical Report of Underwater Archaeological Prospection. Facilities for Nautical Sports in San Antón, La Coruña, 2006–2007). Report to Dirección Xeral de Patrimonio Cultural, Xunta de Galicia, Santiago de Compostela, España, from Arqueología Marítima, A Coruña, ES.

2009   Memoria técnica de la realización de un control arqueológico de dragado en el área objeto de la intervención para la ampliación del Muelle de Bayona, Pontevedra, 2007–2008 (Technical Report of an Archaeological Control of Dredging in the Area Object of the Intervention for the Expansion of the Bayona Pier, Pontevedra, 2007–2008). Report to Dirección Xeral de Patrimonio Cultural, Xunta de Galicia, Santiago de Compostela, España, from Archeonauta, S.L, Oleiros, ES.

SÁÑEZ REGUART, ANTONIO
1793   *Diccionario histórico de los artes de la pesca nacional (Historical Dictionary of National Fishing Gear).* Vda. de Don Joaquín Ibarra, Madrid, ES.

TORRES, C, J. C. CASTRO, AND S. PRIETO
2007   La factoría romana de salazón del yacimiento de O Areal, Vigo (Galicia): un complejo industrial salazonero altoimperial (The Roman Fish-Salting Factory at the O Areal Site, Vigo (Galicia): an Early Empire Fish-salting Complex). In *Salsas y salazones de pescado en Occidente durante la Antigüedad, Actas del Congreso Internacional (Cádiz, 7-9 de noviembre de 2005)*, L. Lagóstena, D. Bernal, and A. Arévalo, editors, pp. 475-486. British Archaeological Reports International Series 1686, Oxford, UK.

VARGAS GIRÓN, JOSÉ MANUEL (EDITOR)
2020   *El instrumental de pesca en el Fretum Gaditanum: Catalogación, análisis tipo-cronológico y comparativa regional (Fishing Instruments in the Fretum Gaditanum: Catalogue, Type-chronological Analysis, and Regional Comparison).* Archaeopress Archaeology, Oxford, UK.

VÁZQUEZ VARELA, JOSÉ MANUEL
2000   *Etnoarqueología: conocer el pasado por medio del presente (Ethnoarchaeology: Knowing the Past through the Present).* Deputación Provincial de Pontevedra, ES.

VILLAR QUINTEIRO, ROSA
2008   Escavación arqueolóxica en área na rúa Tomás Mirambell nº 8, Panxón, Nigrán (Archaeological Excavation in Rúa Tomás Mirambell nº 8, Panxón, Nigrán). *Actuacións Arqueolóxicas* Ano 2006:159–160 <https://issuu.com/davidabella/docs/actuaciones_arqueol__xicas_2006>. Accessed 21 December 2022.

VILLAR QUINTEIRO, ROSA, AND NICOLÁS VILLACIEROS-ROBINEAU
2010   Castro de Panxón (Nigrán, Pontevedra). Nuevos datos y evaluación de su estado actual (Castro de Panxón (Nigrán, Pontevedra): New Data and Evaluation of its Current State). *Gallaecia* 29:137–144 <https://dialnet.unirioja.es/servlet/articulo?codigo=3282763>. Accessed 21 December 2022.

• • • • • • • • • • • • • • •

Laura Casal Fernández
Grupo de Estudos de Arqueoloxía,
Antigüidade e Territorio (GEAAT)
Departamento de Historia, Arte e Xeografía
Universidade de Vigo
Campus As Lagoas s/n Ourense
España CP 32004

Víctor José Barbeito Pose
Adro Arqueolóxica S. L.
Rúa das Barreiras 78 1º Santiago de Compostela
España CP 15702

The Sinking of *Indian* (1817), or How History Resurfaces

Olivia Hulot , René Ogor, Graham Maclachlan (Translation)

On 10 January 1817, at 4 a.m., Indian, *an English three-masted ship of about 500 tons, with 193 people on board, was thrown by a storm onto the reefs of the Kerlouan coast (French Brittany).* Indian *left London under the command of Captain James Davidson, and was part of a fleet of five ships bound for Venezuela, with a contingent of British troops to support Simon Bolivar's revolution. Discovered in 1992 by a sport diver, the wreck was subjected to looting before being formally identified and excavated in 2012. In 2013, more than 300 archaeological objects were the subject of a presentation to the public in the framework of an exhibition. This paper returns to the history of* Indian *through a cross-study of archival documents and archaeological studies, allowing a wider understanding of this wreck.*

No dia 10 de janeiro de 1817, às 4 horas da manhã, o Indian, *um navio inglês de três mastros com cerca de 500 toneladas, com 193 pessoas a bordo, foi atirado por uma tempestade para os recifes da costa de Kerlouan (Bretanha francesa). O* Indian *partiu de Londres, sob o comando do capitão James Davidson, e fazia parte de uma frota de cinco navios com destino à Venezuela, com um contingente de tropas britânicas para apoiar a revolução de Simon Bolívar. Descoberto em 1992 por um mergulhador desportivo, o naufrágio foi sujeito a pilhagens antes de ser formalmente identificado e escavado em 2012. Em 2013, mais de 300 objetos arqueológicos foram apresentados ao público no âmbito de uma exposição. Este artigo retoma a história do* Indian *através de uma investigação cruzada de documentos de arquivo e estudos arqueológicos, permitindo uma compreensão mais alargada deste naufrágio.*

*Le 10 janvier 1817, à 4 heures du matin, l'*Indian, *un trois-mâts anglais d'environ 500 tonneaux, avec 193 personnes à bord, est jeté par une tempête sur les récifs de la côte de Kerlouan (Bretagne française).* Indian *quitta Londres sous le commandement du capitaine James Davidson, et faisait partie d'une flotte de cinq navires à destination du Venezuela, avec un contingent de troupes britanniques pour soutenir la révolution de Simon Bolivar. Découverte en 1992 par un plongeur sportif, l'épave a fait l'objet d'un pillage avant d'être formellement identifiée et fouillée en 2012. En 2013, plus de 300 objets archéologiques ont fait l'objet d'une présentation au public dans le cadre d'une exposition. Cet article revient sur l'histoire de l'*Indian *à travers une étude croisée de documents d'archives et d'études archéologiques, permettant une compréhension plus large de cette épave.*

Site Location and Discovery

Today, in France, material remains lying in the depths of the sea are protected by law, under the *code du patrimoine* or 'heritage code'. Since 1966, a special department of France's Ministry of Culture, the Département des Recherches Archéologiques Subaquatiques et Sous-marines (Department of Underwater Archaeological Research), or DRASSM for short, is tasked with cataloging, studying, protecting, and promoting this underwater cultural heritage.

The wreck of *Indian* lies in the English Channel, in waters under French jurisdiction, and more precisely off the north coast of Finistère, an area off French Brittany. This jagged coast is all the more inhospitable because of strong currents, frequent storms, and the many rocky shoals that make navigation in the area hazardous. Sometimes called in French *la côte des légendes* (coasts of the legends) or *les Pays pagan* (Pagans

countries), this portion of the coast has a reputation for being home to 'wreckers,' who plundered stricken vessels and might have even caused their ruin. This reputation, although greatly exaggerated, does nevertheless attest to the opportunity afforded to coastal populations, during some historical periods, for salvaging equipment and foodstuffs from ships wrecked on the coast.

In 1991, the remains of a wreck, at that time unidentified, were officially discovered and registered with the French authorities. Subsequent historical investigations revealed it to be that of *Indian*, a three-masted British ship of approximately 300 tons, which foundered off the French Brittany coast in the early morning darkness of the 10 December 1817 (Figure 1). Despite being situated in shallow waters of between 6 and 10 meters (m), the site of *Indian* was not surveyed immediately, and plunderers paid it several visits. (Figure 2). In addition, the remains have steadily deteriorated because of erosion and the voracious appetite of xylophagous mollusks (Loiselet 1991).

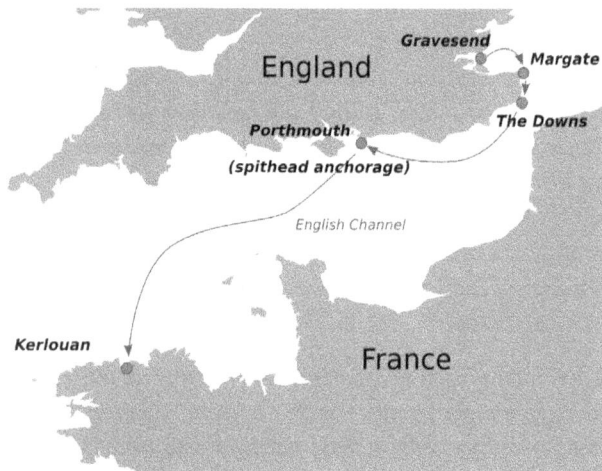

FIGURE 1. Course of *Indian* from Gravesend (UK) to the shipwreck site at Kerlouan (French Brittany). (Map by the primary author, 2023.)

FIGURE 2: Bronze bell from *Indian*, recovered after looting. (Photo by the primary author, 2015.)

Identifying *Indian*

In 2011, René Ogor (volunteer diver and historian) began to identify artifacts salvaged illegally from the site and set about scouring the records for information,

clocking up almost 300 hours of research. A first dive on the wreck revealed a few artifacts that provided information on the date and probable nationality of the shipwreck (as recorded by Ogor in 2011). This included:

- An earthenware shard bearing a poem. A poem in the English language, by an anonymous poet from the south of England and dated to the first half of the 18th century.

- Several white clay pipe stems. The diameter of the pipes indicate they were made after 1775.

- A bugle mouthpiece.

- Gun flints that are indicative of what the English army used after 1816.

- Intrigued and curious to know more, René Ogor met with the original discoverer of the site, who had removed artifacts at the time of discovery. Some of these objects helped reveal the identity of the wreck, including: bronze spur holders; two, gold, marine chronometer bottoms, identical to those recovered from the wreck of the *Belona*, an English merchant ship that wrecked in 1814 on the Talbert Trench; and one cast iron gun, engraved 1804.

But ultimately, it was a uniform button bearing the inscription *1 HUSS VENEZUELA* (Figure 3), with an iron carronade bearing the date of 1804, and the presence of British earthenware which consolidated the identification of the site. The military history of the Napoleonic Wars ended with Napoleon's defeat at Waterloo and his relegation to the island of St. Helena. In the two years following the Battle of Waterloo, nearly 500,000 British soldiers were demobilized. Returning to Britain, most of these men found themselves facing an uncertain future, even poverty. At the same time, Simon Bolivar, who had decided to liberate Latin America from the Spanish invader, in April 1817 sent his agent, Luis Lopez Mendez, to secretly question the British Foreign Office on its official position concerning the recruitment of an army of British legionaries (O'Leary 1969). It would seem that the British government saw this as a way to alleviate the problem of demobilization.

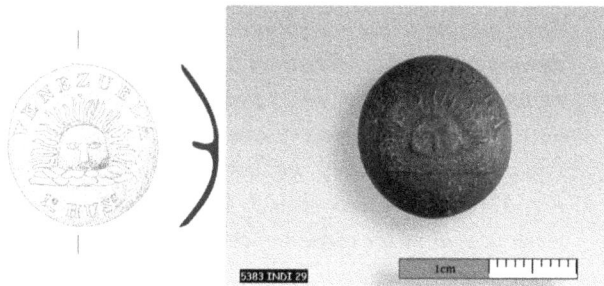

FIGURE 3. Uniform button bearing the inscription. (Photo by R. Ogor 2013; Drawing by MN. Baudrand 2015.)

At the end of May 1817, Luis Lopez Mendes, who had set up shop at 27 Grafton Way, London, began enlisting troops. Volunteers were encouraged by promises of pay equivalent to that of the British army and promotion to a higher rank than they previously held. Thousands of volunteers came forward, Irish and English, and the first regiments took shape. The British government, hopeful of a solution to the problems of demobilizing its soldiers, and seeing an opportunity to secure a new market for its exports, agreed to provide assistance to the revolutionaries. Five ships were mobilized for this expedition (O'Leary 1969).

- *Indian*: a ship of 508 tons, Captain Davidson, owner William Gibbon, carried the Venezuelan 1st Lancers of Colonel Robert Skeene (about 220 people). It carried most of the equipment, and the pay of the cavalry regiments (Hippisley 1819).

- *Prince*: a light frigate of about 400 tons, Captain Nightingale, carried Colonel Henry Wilson's 2nd Venezuela Hussars. Major Graham had his wife on board. One of the first to join the ranks of the 2nd Venezuelan Hussars, Daniel Florence O'Leary was promoted to the rank of general, after many years of service with Simon Bolivar. His memoirs, *The Detached Recollection of General D. F O'Leary*, published in 1969, by the Institute of Latin America Studies of the University of London, are now recognized as one of the most important sources for the study of the South American liberation campaigns.

- *Dowson*: a ship of about 400 tons, Captain Dormor, carried the Venezuelan 1st Rifles under Colonel Donald Campbell's command (Hippisley 1819).

- *Emerald*: a ship of about 460 tons, Captain Weartherley, carried the 1st Venezuelan Hussar of Colonel Gustavus Hippisley, who described his Bolivian campaign in a book entitled, *A Narrative of the Expedition to the Rivers Orinoco and Apure in South America: Which Sailed from England in November 1817, and Joined the Patriotic Forces in Venezuela and Caraccas* (Hippisley 1819). It provides particularly precise information on the preparations for the expedition and the composition of the expeditionary corps.

- *Britannia*: a vessel of about 400 tons, commanded by Captain Sharpe, in charge of carrying artillery (Hippisley 1819).

Five detachments were established, and these comprised two regiments of Venezuelan Hussars, one of riflemen and another one of lancers, and a brigade of artillery (Hippisley 1819).

- *1st Venezuelan Hussars*: 30 officers and 160 noncommissioned officers, commanded by Colonel Gustavus Hippisley.

- *2nd Venezuelan Hussars (Red Hussars)*: 20 officers and 100 noncommissioned officers, commanded by Colonel Robert Skeene.

- *1st Venezuelan Lancers*: 20 officers and 200 noncommissioned officers, commanded by Colonel Henry Wilson.

- *1st Venezuelan Fusiliers*: 37 officers and 200 noncommissioned officers, commanded by Colonel Donald Campbell.

- *Artillery Brigade*: 10 officers and 80 noncommissioned officers, commanded by Colonel Joseph. A Gilmore

Each regiment had its own equipment, conceived, and designed especially for this expedition (Hippisley 1819). *Indian* was carrying the 2nd regiment of Venezuelan Hussars commanded by Colonel Robert Skeene. A total of 193 people were aboard, including

150 soldiers, 24 members of the ship's crew, 12 women, and one child. All the military equipment for the expedition was supplied by London merchants, Thompson and Mackintosh. (Hippisley 1819).

Everything should have been ready by August 1817, but given the quantities involved, the delivery was delayed, and the ship's departure dates were pushed back by several months. A large quantity of military equipment, clothes, and a few men from the same corps, the buglers Hodge, Spearman, and Stevenson; the veterinary surgeon, Powis; three corporals; three sergeants, and two veterinarians, were moved aboard *Indian* and *Prince*, the ships with the greatest tonnage. The uniforms belonging to the 1st Venezuelan Hussars commanded by Colonel Hippisley were, therefore, on board *Indian* (Hippisley 1819).

Indian finally set out from the English coast on the 2 December 1817 under the command of Captain James Davidson. Despite the bad weather, the captain was ordered to get underway because the ship was at risk of being detained in port following complaints made through diplomatic channels by the Spanish government. At around four o'clock in the morning on 10 December 1817, *Indian* was driven onto the coast. Witnesses recalled the storm, which lasted several days, was exceptionally violent. There were no survivors. At first light, the locals discovered the scale of the disaster. The beach was littered with bodies, goods, and parts of the ship's rigging. A couple of pigs were the only living things to have escaped the tragedy, having washed ashore by some miracle in one of the ship's boats. Jacques Boucher de Perthes who was, in 1817, Director of Customs in Morlaix and responsible for the sector, published his arrival on the site of the wreck in his memoirs entitled *Under Ten Kings - Memories of 1791 to 1860*:

[...] "This misfortune, however great it was, was nothing compared to what was happening further on. An English three-masted ship, named the India, which carried to Colombia recruits for the cavalry and some other passengers of both sexes, in all four hundred and some people, as we learned since then, perished at this very moment a few leagues from us, on the coast of Plouguarneau.

I ran there at full speed, but the disaster was over. Nothing had escaped : corpses everywhere. I could not take a step without my horse being stopped by a figure of a man with a moustache.

From far and wide, one encountered the body of a woman or a child. More than twenty times I dismounted, more than twenty times I put my hand on a chest and brought my ear close to a mouth. Not a heart was beating, not a mouth was breathing. Many bodies were even disfigured by wounds received against the corners of the rocks and debris of the ship.

When I arrived at the place where the India had broken up, this vessel, which must have been seven or eight hundred tons, seemed to have been chopped up. Her masts were floating on one side; on the other, her yards and topsails. Her rudder was dry, along with part of her bow. Here again, trunks, bundles, and chests of all shapes were floating, clashing, forming a vast moving ruin.

"On the shore, open boxes, many of them filled with military effects. What was most noticeable were the woollen caps with the three colours : red, blue and white. Hundreds of peasants, men, women and children, had already put on their caps. These caps, whose shape was that of our cotton caps, gave to these figures an aspect that one would have found laughable in any other circumstance.

A strange thing, which can be explained by the conformation and the strength of will of certain animals, is that out of the four hundred or so people on board, when not one of them ran away, a dozen pigs arrived on land safe and sound.

The number of ships wrecked during the night on the left side of my division, from Roscoff to Abreuvack (Aberwrac'h), or over an area of twelve leagues, is six.

For a week, the coast was covered with corpses; the sea was constantly bringing them back. They were poor soldiers, and there was no hurry to bury them: it was a job and the people here, no more than elsewhere, do not like to work for nothing. And they were heretics."[...] (Boucher de Crèvecœur de Perthes 1863)

While the location of the burial site of the 143 recovered corpses is not known, the action of the sea has, in recent years, uncovered human remains at the foot of the dune, not far from the wreck site. These could belong to the passengers and crew of *Indian*. In 2012 and 2013, an archaeological project began, starting with an assessment, followed by a magnetometer survey, and then a series of test excavations (Figure 4).

First Field Investigations

The main part of the site consists of the keel, a few meters long, which lies between sand and rocks on the side of a rocky cliff, at a depth of 7 to 10 m at low tide, in an area sheltered from the currents (but not from the swell). The magnetic signature of

FIGURE 4. Archaeologist working on the site. (Photo taken by T. Seguin, 2013.)

the rocks prevented the survey from locating guns or devices. The test excavations revealed the position of the aft section of the ship, which was lying close to the coast. Little sedimentation was present, and the rocky substrate was not deep. Artifacts survived in small sand-filled cavities but were fragmentary. A series of hussar sabers were discovered at the foot of the main

rocky outcrop. It was an isolated find, which suggests that a crate had drifted from the wreck before being rapidly buried under sediment. The ship's bell bears the number '1810,' the year the vessel was launched. It was discovered by a spearfisher some 600 m to the northwest of the aft section of the wreck.

Heavy sedimentation was present in this sector and no trace of the forward section of the vessel was found during the archaeological investigations. An article published in *The Times* (1818) mentioned that the bow had been pulverized and that only the stern remained intact, its upper section being visible at low tide. If artifacts have survived in the sediment, it's likely that they are well preserved and safe from collectors.

In 2012, as the archaeological investigations progressed, many artifacts plundered from the site were recovered, while others were discovered during field operations. These artifacts contributed to a better understanding of the ship's equipment and the daily life aboard. *Indian* was privately owned, measured 35 to 40 m in length, and probably carried 12 to 24 guns. Only two carronades were spotted on the site. It is possible that some pieces of artillery were salvaged after the wrecking or are still on the site, lying hidden in the sediment. The carronade brought up from the site is marked with a royal crown depicted on a shield, its caliber is 18 pounds, and its date of manufacture is 1804.

The artifacts found in the stern section originate from the officers' quarters, as attested by the discovery of an hourglass, a piece of a telescope, and the backs of two gold chronometers. On one of these, the inscriptions indicate that it was made from 18-carat gold between 1811 and 1812 by Margas Jacob working for the Goldsmiths Company. The other, also in 18-carat gold, was made by Benjamin Smith working for the same company in around 1810. It bears other initials that could indicate its owner.

The list of passengers sailing aboard *Indian* was published in *The Times* on 17 December 1818 (*The Times* 1818) and mentioned the names of the regiments' three buglers: Little, Dutton and Read. Two mouthpieces were found at the site, as was a bugle with its bell missing. Many remains related to military equipment were found on the site, such as items for decorating uniforms, epaulettes, a powder flask, numerous gun flints, spurs, and a series of buttons belonging to the uniform of the 1st Venezuelan Hussars. Daily life aboard *Indian* can be glimpsed through the many remains of tableware, including white and decorated earthenware, glass bottles, and tin plates. Some bear the initials of their owner, subtly inscribed on the back. The initials 'WRS' suggest the bugler, William R. Spearman. Other remains, such as an inkwell, toothbrushes, hairbrushes, and clay smoking pipes, evoke the daily lives and habits of the officers, the men-at-arms, and the sailors aboard *Indian* before the final toll of the ship's bell on that cold December night in 1817.

Public Outreach

This fascinating page of maritime history was presented to the general public through a touring exhibition, which featured artifacts recovered from the wreck. The exhibition was hosted at four separate museums. By the end of summer 2013, the exibition was seen by over 20,000 people in the space of three months. Subsequent publications may further address the intervening history of this unique maritime site.

References

HIPPISLEY, GUSTAVUS
1819   *A Narrative of the Expedition to the Rivers Orinoco and Apuré in South America: Which Sailed from England in November 1817, and Joined the Patriotic Forces in Venezuela and Caraccas.* Murray, London, UK.

BOUCHER DE CRÈVECŒUR DE PERTHES, JACQUES
1863   *Sous dix rois: Souvenirs de 1791 à 1860 (Under Ten Kings: Memories from 1791 to 1860).* Jung-Treuttle, Paris, FR.

LOISELET, M.
1991   Maritime wreck declaration. Registered with DRASSM (Departement des recherches archéologiques subaquatiques et sous marine), FR.

O'LEARY, DANIEL FLORENCIO
1969   *The Detached Recollections of General D.F. O'Leary.* R.A. Humphreys, Institute of Latin America Studies, London, UK.

THE TIMES
1818   *Crew of the Indian.* 17 February 1818. London, UK.

· · · · · · · · · · · · · · ·

Olivia Hulot
Department of Underwater Archaeological Research
French Ministry of Culture and Communication
147 Plage de L'Estaque
13016 Marseille
France

René Ogor
16 rue Sully Prudhomme
29200 Brest
France

Graham Maclachlan (Translation)
6 rue de la chapelle
29340 Riec-sur-Bélon
France

The Port de Pomègues 4 Wreck: A Lead-Sheathed Ibero-Atlantic Vessel

Marine Jaouen, Sébastien Berthaut-Clarac

The site of the Port de Pomègues 4 wreck is characterized by the port side of a ship estimated to be about 23 meters long. The remains of this oak hull, with lead sheathing, lie at a depth of four meters on the northern coast of the island of Pomègues in the Bay of Marseille (France). The absence of artifacts and dendrochronological results do not suggest an absolute date. The presence and analysis of lead sheathing, indicates a chronological range between the 16th and the early 17th centuries. In addition, the hull has some Iberian-Atlantic shipbuilding signatures.

O naufrágio Port de Pomègues 4 preserva o lado bombordo do navio com um comprimento estimado em cerca de 23 metros. O vestígio do casco, em carvalho, com revestimento de chumbo, encontra-se a uma profundidade de quatro metros na costa norte da ilha de Pomègues, na baía de Marselha (França). A ausência de artefactos e os resultados dendrocronológicos não sugerem uma data absoluta. A presença e a análise do revestimento de chumbo indicam um intervalo cronológico entre o século XVI e o início do século XVII. Para além disso, o casco apresenta algumas características da construção naval ibérico-atlântica.

Le site de l'épave du Port de Pomègues 4 est caractérisé par le côté bâbord d'un navire estimé à environ 23 mètres de long. Les restes de cette coque de chêne, avec un revêtement de plomb, se trouvent à une profondeur de quatre mètres sur la côte nord de l'île de Pomègues dans la baie de Marseille (France). L'absence d'artefacts et de résultats dendrochronologiques ne suggère pas une date absolue. La présence et l'analyse du revêtement en plomb indiquent une plage chronologique entre le 16ème et le début du 17ème siècle. En outre, la coque a quelques signatures de construction navale ibéro-atlantique.

Introduction

The Frioul archipelago, in the harbor of Marseille, France, is composed of two large islands, Ratonneau to the north and Pomègues to the south. This archipelago, occupied since antiquity, was the outpost of the port of Marseille, founded by the Greeks six hundred years before present. In the 15th century, strategic secondary ports on the islands of Endoume, Ratonneau, and Pomègues completed the main dock, now called the Vieux-Port (Figure 1). Indeed, ships of more than three hundred tons could not dock in the Vieux-Port of Marseille because of the natural accumulation of silt and the practice of unlawful dumping of ballast. Moreover, the possession of these islands ensured the maritime traffic control of Marseille.

In 1977, Marie-Antoinette Marcos and Serge Ximénès found and declared the wreck of Port de Pomègues 4 (PP4). Later, it was the subject of surveys in 2013 under the leadership of Michel Goury. From 2019 to 2021, Marine Jaouen conducted excavations on behalf of the French Ministry of Culture. PP4 is located at a depth of 4 meters (m), at the bottom of a natural cove sheltered from the prevailing wind and at the tip of a spit of land, which in the 17th century served as a harbor master's office in the port of Pomègues. Exogenous river pebbles litter the seabed. They probably came from the ballast of quarantined ships.

The site is characterized by a portion of the port side of a ship, built entirely of deciduous oak and sheathed with lead. In total, 18 m of hull are preserved from the aft part of the stern curve to a few meters after the main beam. Transversely, the remains start with the starboard side and are preserved up to the first port side extensions, whose heads are degraded by *Teredo navalis* (shipworm). The maximum width is 5 m. (Figure 2). Surprisingly, no artifacts were found on the wreck during the three years of excavations. Furthermore, the dendrochronological samples did not produce a date for the felling of the wood because trees were cut too young to be dated.

Characteristics of the Wreck

The preserved hull portions include 14 rows of hull planking/carvel-built and plain scarf (45 to 50 centimeters (cm) wide, 6 to 7 cm thick); six rows of planks are distributed from the keelson to the floor-heads (20 to 40 cm wide and 7 to 8 cm thick). The floor-head ceiling is cut to receive them. Fifteen floor-heads are preserved; they are precisely fitted into the space. These elements constitute the preserved longitudinal structure.

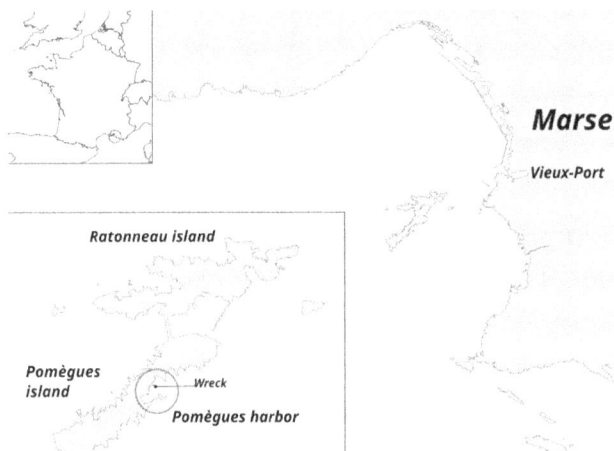

FIGURE 1. Location of Port de Pomègues 4, true north. (Figure created by the primary author, 2023.)

Nine of the 13 crotches are preserved; five of them are still connected to their futtock and were fitted to the stern curve, as well as joined by pins. The midship frame was identified during the planimetric survey. By identifying each floor timber with a letter/number designation, VR87 appeared to have two futtocks. This identification is confirmed by the orientation of the bevels on the foot of the futtocks, which also change between futtocks All87 and All85 (Figure 3). Twenty-six floor timbers, with their futtocks attached, are present, which together with the crotches constitute the preserved transverse structure of this wreck.

Transverse Structures

The width of the floor timbers is between 18 and 20 cm, like the space between them. Floors and futtocks have the same width. At the junction between floor timbers and futtocks, the space disappears, reinforcing

the hull by an uninterrupted succession of floor/futtock/floor. The floor timbers are joined to the keel without lateral overlapping. The limber hole is single, central, and rectangular (6 × 7 cm). Pin marks are visible on the lower face of the keel. The assembly from the outside of the keel to the inside of the hull required a particular process by the shipyard. More traditionally, the planks are fastened to the floors from the outside to the inside.

Floor timbers VR83 and VR84 are cut in such a way as to produce a notch of circular section (35 cm in diameter), which could correspond, given its location at the back of the mid-ship frame, to the foot of a bilge pump. Between these two floor timbers, a small piece of wood is placed on the planking and reinforces the location of a pillar. In addition, two other arrangements of rectangular sections (20 × 16 cm) are also interpreted as being linked to pillars. The lower part of the futtocks is beveled from timbers All97 to All91. The bevel is again visible but reversed between futtocks All84 and All78. The other lower parts of the futtocks are either too degraded to be described or are hidden under the ceiling. The tipping point of the orientation of the bevels is located on the midship frame. The assembly floor/futtock is a rectangular dovetail (8.7 × 2.0 cm) (Figure 4a and 4b). On PP4, the floor and its futtocks are fastened by at least one nail that passes through both pieces at the bevel of the futtock foot. The density of the construction did not allow any observation. No futtock head is preserved, and the planking disappears beyond the upper limit of preservation of the first futtock.

The last four crotches have disappeared. The sternpost knee was cut to accommodate them. On this part of the hull, traces of the pitch are clearly visible and have

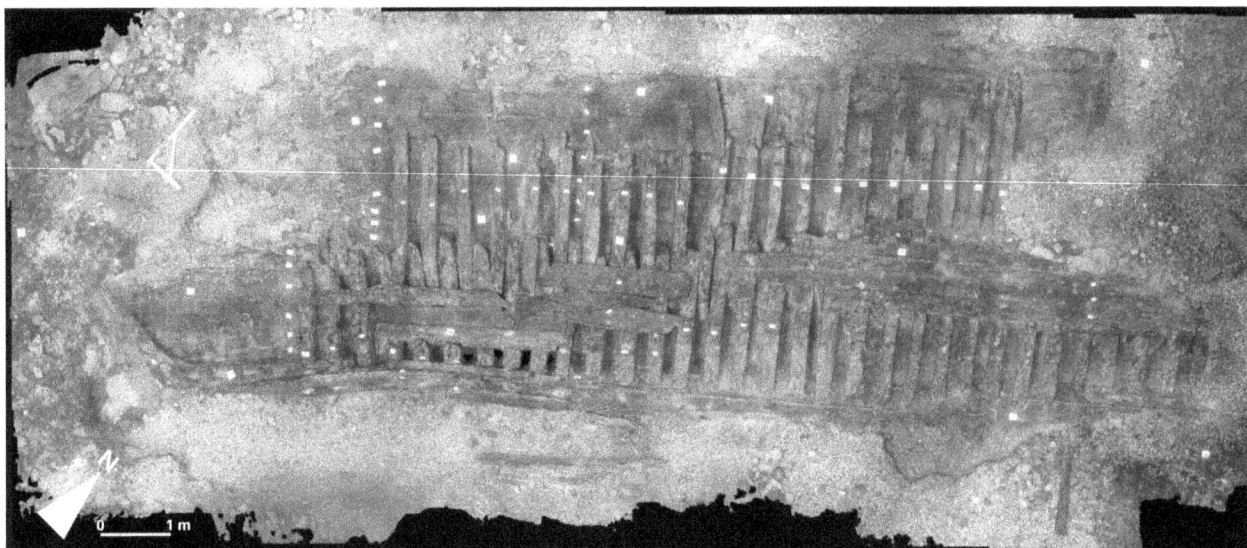

FIGURE 2. Orthophotography of the PP4 wreck, true north. (Figure created by Teddy Seguin, MC/DRASSM 2021.)

Advisory Council on Underwater Archaeology

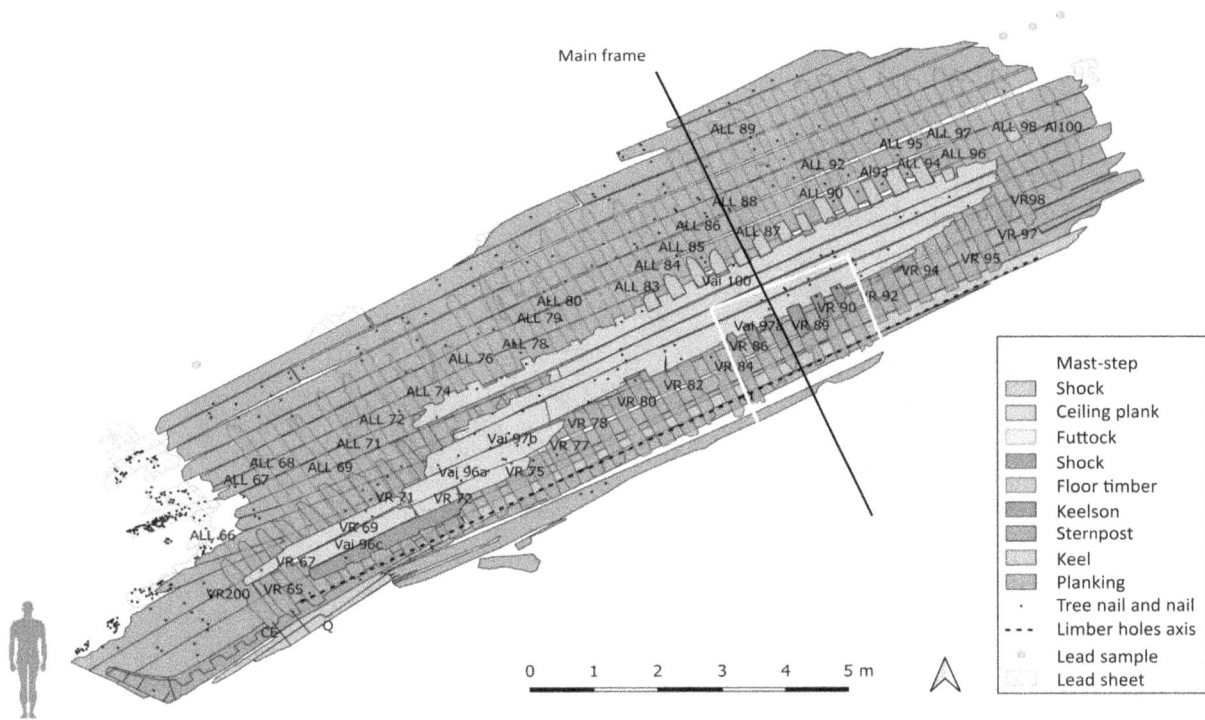

FIGURE 3. General planimetry of the preserved remains, true north. (Figure created by the primary author, 2021.)

contributed to the watertightness of the hull bottom. The foot of the crotches is positioned on each lateral sides of the sternpost knee. The following crotches are positioned on the keel.

Longitudinal Structures

The ends of the keel are missing, but the horizontal part of the sternpost knee is entirely preserved. Its angle is 60 degrees. The keel is composed of three elements, and the upper face is 24 cm wide and slightly convex. From the back to the front, the first section is located under the sternpost knee; its height is 23 cm. The height of the second section is 20 cm. A third section starts is between VR72 and VR73. Here the lap joint is scarf on the upper side of the keel (Figure 4c). On midship frame, the height of the keel is 17 cm.

The keelson is not complete. Its width is 18 cm, its height is between 10 and 22 cm. Its upper side is flat. Its lower face is cut to fit on the crotches. On its upper surface, a hollowed-out arrangement allows the positioning of a pillar (60.0 cm long, 8.5 cm wide, 2.0 cm deep). Only VR67 seems connected to the keelson by a vertically inserted pin and, thus, marks the limit between the floor timber and the crotches.

From their central location, the preserved crotches are associated with the main mast. The six crutches rest on

six floor timbers (VR90 to VR85) and end astride the ceiling Vai97. The complete shape of these six crutches is not known due to the action of woodborers. The identification of the midship frame upstream of the crutches seems to confirm this hypothesis. This one is slightly in front of the main mast. These two factors allow the suggestion of a probable hull length of around 23 meters.

Floor-heads, inserted in force, fill the gap between two futtocks. The first ceiling preserved is the rung-head ceiling; it is cut to accommodate each floor-head. Only 16 are preserved. Between two floor-heads and on the upper side of the futtock, a triangular section piece completes the device. These pieces are considered the "false floor-head." Seven of these pieces are still in place.

The ceiling is preserved on seven rows from the floor-head ceiling to the keelson. All the ceiling planks have not been preserved. Nevertheless, it was determined that no more than seven rows of ceiling existed. The ceiling planks are fastened with a plain scarf or by a beveled scarf. One of them (Vai99) is not flat, but cut in a half-trunk shape. It serves as a longitudinal reinforcement in the manner of a bilge stringer.

The width of planks is large, between 45 and 50 cm. Their thickness is between 6 and 7 cm. The dismantling of four planks provided a view of the work done to avoid that the heads of the nails fixing the strakes on the

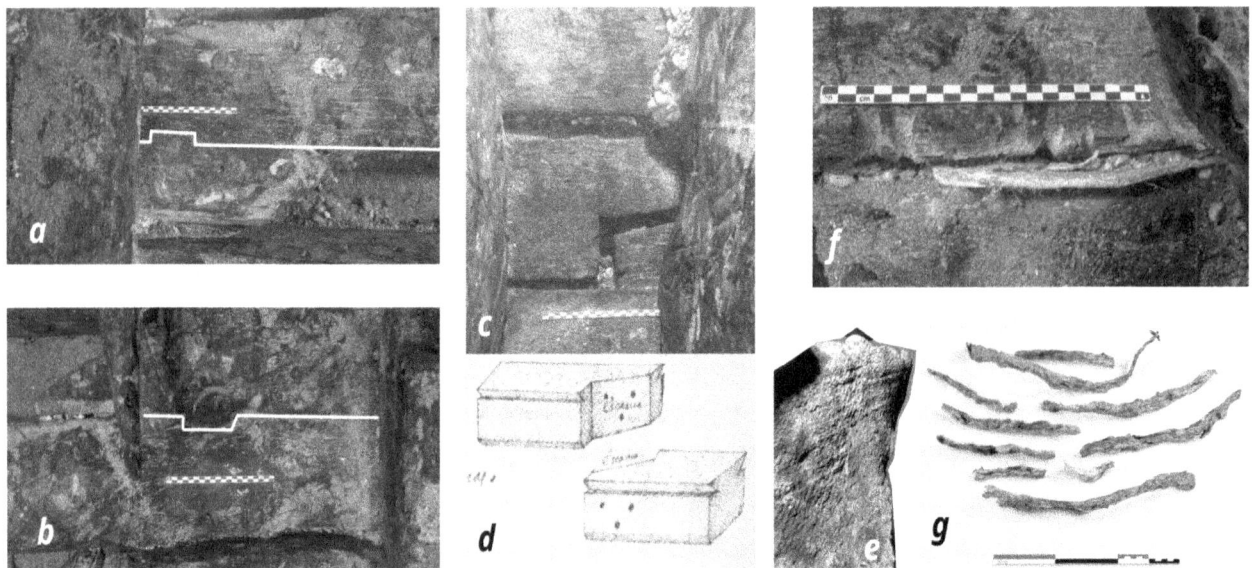

FIGURE 4. Characteristic elements of the Port de Pomègues 4 wreck (*a*) and (*b*) dove-tail gap at the junction of the floors with the futtocks (*c*) keel junction (*d*) drawing by João Baptista Lavanha, 1608, keel junction described in "O Livro Primeiro de Architectura Nava", f.62, (*e*) traces of tissue on the lead sheathing (*f*) lead-string in place and (*g*) lead-string after recovery. (Figure created by Teddy Seguin, MC/DRASSM 2021.)

frames exceed the thickness of the planking. This precaution had two advantages. It maintains the hull without roughness. It allows the heads of the nails, probably made of iron, not to be in direct contact with the lead sheets. This precaution limits electrolysis and punching of the lead sheets. The pins have a square section (1.0 to 2.0 cm) and a round head (3.2 cm in diameter). With their removal, deep striations have been observed on the external face of planks. For wood specialists, this is not the result of the action of woodborers but perhaps the sign of a long drying of the planks in the open air.

Lead Sheathing

The outer planking of this wreck was covered with lead sheets, the largest preserved pieces of which measure 182.0 cm × 42.0 cm, with a thickness of 1.0 to 1.5 millimeters (mm). Round-headed iron nails, with square shanks, were used to secure these lead sheets, spaced 4 cm apart in all directions without rigorous alignment. The outer surface of the planking does not show evidence of successive reinstallations of the sheathing. The observed sheathing is, therefore, in all probability the first one made on the ship. Lead sheets have traces of fabric on their inner surface attributable to the table-cast fabrication process with fabric (Figure 4e). The interface between the sheets and the planking was analyzed and found to be dry pitch, pitch with no added animal material. The keel is neither fitted with a false keel nor

lead-sheathing. The first sheets of lead are nailed to the garboard.

Lead is also employed as a punctual caulking element. Sheets of lead folded on themselves were pushed horizontally between two planks (Figures 4f). These sheets are all the same width, only their length changes depending on the area to caulk (Figure 4g). These lead sheets do not have the same thickness as the hull sheathing. They are thinner, with a thickness of between 0.4 and 0.8 mm.

An Iberian-Atlantic Shipbuilding Tradition?

In her article, Loureiro defines, following Oertling (2001), 14 architectural signatures of Iberian-Atlantic shipbuilding tradition (Loureiro 2012). Port de Pomègues 4 has nine of these signatures (Table 1). For Oertling, the keelson as described above is one of the reliable signatures defining this tradition (Oertling 2001:237).

In addition, many points refer to similarities of wrecks of Iberian-Atlantic shipbuilding tradition, notably the wreck of *Nossa Senhora dos Mártires*: the keel scarf, central limber hole, assembly by tenons and mortises, assembled floor and futtock and "lead string" (Castro 2003a). But the dimensions of the ship, Portuguese Indiaman launched in 1605 and lost in 1606, are much larger than that of PP4: three masts, 50 m overall, 13 m wide.

The keel scarf is known on the wrecks Aveiro A, from the 16th century; Highborn Cay, first quarter of the 16th century (Oertling 2007); and *San Juan*, 16th century (Alves et al. 2001). An illustration of this type of assemblage is given in Lavanha's 1608 book, *O Livro Primeiro de Architectura Naval* (First Book of Naval Architecture).

The central limber hole is attested to on the wrecks of Arade 1 and *San Juan*, whose flat floors as on PP4 aren't assembled on lateral sides of the keel (Loewen 1998). Moreover, the floors and the futtocks are assembled by tenons and mortises, which adds an additional similarity. It is worth mentioning the Arade 1 site, whose floors have the same specificities (Loureiro 2016), and the 16th-century Western Ledge Reef wreck (Bojakowski 2011), which has similar configurations. The embedding of the ceiling is specific to this tradition and bears the specific name of *albaola* in the Basque language. The wreck Arade 1 has a similar type of floor-head and "false floor-heads" (Castro 2003b), and only the rung-head ceiling differs.

Regarding the planking, the presence of the same striations is found on the wrecks of Angra D, a ship of the late 16th and early 17th centuries and *La Nossa Senhora dos Mártires* (Castro 2003a). Such traces of fabric could be observed on the sheathing or lead elements of the Spanish wrecks of Padre Island, dated 1554, Angra D (Garcia and Monteiro 1998) and *Santa Margarita*, wrecked in 1622 (Malcolm 2001), but also on the Genoese wreck of *San Juan*, dated 1581 (Ridella et al. 2018).

Port de Pomègues 4 and the Lead-Sheathed Ships

The lead sheathing is known in antiquity, but seems to disappear gradually from the second century A.D. (Kahanov 1999:219). This technique reappeared in the 16th century. Sixty lead-sheathed ships were identified in the literature between 1513 and 1833, including 13 wrecks (PP4 was part of this discovery). Of these 60 ships, 45% were English; 13% Spanish; 8% Genoese, Dutch, or French; 5% were Ragusan or unknown; and less than 2% were Maltese or Venetian. Only 11 ships were built in the Mediterranean.

Architectural Signatures of Iberian-Atlantic Shipbuilding Tradition (Loureiro, 2012)	Port Pomègues 4
A predefined proportional play between the key dimensions of the ship: depth, extreme breadth, overall length and keel	?
The "one full, one empty" frame	Yes
The middle region of the transverse frame designed using predefined diagrams	?
The central frames mounted before their installation on the keel	Yes
The floors timbers at the ends of the ship defined with the help of ribband	?
The planking, carvel-built , installed after the installation of the central frames	Yes
The keelson, stringers and wale as longitudinal reinforcement elements	Yes
Gaps and tenons as a fastening system for certain structural elements or sections of the same part	Yes
The keel skeg as a connecting element between the keel and the sternpost	?
The ceiling covering the surface of the floor timber	Yes
The keelson fitted on the upper face of the floors by notches dug in its lower surface	Yes
The mast step of the main mast integrated in the keelson and the sump intended to receive the pump tube cut, in whole or in part, in this element	Yes
The mast step of the mast surrounded on the sides by cleats	Yes
The flat transom completing the sternpost	?

TABLE 1. Characteristics of Iberian-Atlantic shipbuilding tradition found in Port de Pomègues 4. (Table created by the authors, 2023.)

In chronological terms, of the 47 ships the exact date of sinking is known or have sheathing, 34% date from the 16th century, 57% from the 17th century, 6% from the 18th, and 2% from the 19th century. This indicates more than 91% of these ships were sheathed in the 16th or 17th century. It is, therefore, reasonable to place Port de Pomègues 4 in this chronology.

Conclusion - Towards the Identification of PP4: *Nicholosa*?

The characteristics of these remains suggest that the dimensions of the ship were of the order of 23.0 m long, 7.5 m broad, with a draft of 3.3 m. From these estimated dimensions, a load capacity of about 108 tons is suggested. Its characteristics indicate that it is of an Iberian-Atlantic shipbuilding tradition, probably from the 16th or 17th century.

French archives mention Iberian maritime trade to Marseille since the modern age. It seems that the cargoes were composed in part of iron, wool, fabrics from Segovia, as well as oil, cereals, and fish. However, to this day, no shipwrecks of Spanish or Portuguese construction has been found near Marseille, while more than 50 wrecks dated between 1492 and 1789 are known in the region. Port de Pomègues 4 seems to be the first archaeological evidence of Iberian navigation to Marseille in the modern age.

Research conducted by Philippe Rigaud (historian and researcher associated with the Laboratory of Medieval and Modern Mediterranean Archaeology) has led to a possible identification. In 1535, Martines Sans de Miranda from San Sebastian (Spain), owner of a 3,000-quintals nautical gauge (a 16th century Marseille measure) called the *Nicholosa*, sold her ship which had been wrecked by a storm in the port of Gallinia or Galliana (Port Pomègues?). Five *maîtres de hache* (carpenters) from Marseille acquired it for the purpose of exploiting the woods (Archives Départementale des Bouches-du-Rhône 1535:f 236v to 241).

Acknowledgments

To Michel Goury, who led us to this wreck and to the volunteer divers and the Calanques National Park divers. To the Ministry of Culture, which has supported these operations financially and logistically. We would like to thank Eric Rieth for the discussions we had about this site, and Pierre Adam and Philippe Schaeffer from the Institute of Chemistry of Strasbourg, for their

analysis of the pitch. We also thank Frédéric Guibal from the Mediterranean Institute of Biodiversity and Ecology Laboratory for his attempts at dating using dendrochronology and Philippe Rigaud for his research in the archives and the transcriptions.

References

ARCHIVES DÉPARTEMENTALE DES BOUCHES-DU-RHÔNE
1535   Notaire, Étude Maubé (Marseille), Fond Perrin, 356 E 305, f°236v to 241.

ALVES, FRANCISCO, PAULO RODRIGUES, MIGUEL ALELUIA, RICARDO RODRIGO, CATARINA GARCIA, ERIC RIETH, AND EDUARDO RICCARDI
2001   Ria de Aveiro A: A Shipwreck from Portugal Dating to the Mid-15th Century; a Preliminary Report. *The International Journal of Nautical Archaeology* 30(1):12–36.

BOJAKOWSKI, PIOTR
2011   The Western Ledge Reef Wreck: Continuing Research on the Late 16th-/Early 17th-century Iberian Shipwreck from Bermuda. P*ost-Medieval Archaeology* 45(1):18–40.

CASTRO, FILIPE
2003a   The Pepper Wreck, an Early 17th-Century Portuguese Indiaman at the Mouth of the Tagus River, Portugal. *The International Journal of Nautical Archaeology* 32(1):6-23.

2003b   *The Arade 1 Ship - 2002 Field Season. Vol. 2: The Hull, ShipLab Report 5.* On file in IPA/CNANS' library, and Nautical Archaeology Program Library, Texas A&M University. <https://www.academia.edu/2026708/The_Arade_1_Ship_2002_Field_Season_Vol_2_The_Hull_ShipLab_Report_5>, Accessed: 30 November 2022.

GARCIA, CATARINA, AND PAULO MONTEIRO
1998   The Excavation and Dismantling of Angra D, a Probable Iberian Seagoing Ship, Angra Bay, Terceira Island, Azores, Portugal. Preliminary Assessment. In *Trabalhos de Arqueologia 18 - Proceedings*, Francisco Alves Editor, pp. 413-47. Lisbone. <http://www.patrimoniocultural.gov.pt/en/publications/trabalhos-de-arqueologia-18-proceedings-international-symposium-on-archaeology-of-medieval-an-modern-ships-of-iberian-atlantictradition-hull-remains-manuscripts-and-ethnographic-sources-a-compar>, Accessed: 30 November 2022.

KAHANOV, YAACOV
1999   Some Aspects of Lead Sheathing in Ancient Ship Construction. In *5th International Symposium on Ship Construction in Antiquity*, Tropis V, Nauplia, 26–28 August 1993, 219–224. Athens: Hellenic Institute for the Preservation of Nautical Tradition.

Lavanha, João Baptista
1608   O Livro Primeiro da Architectura Naval (First Book
        of Naval Architecture). Translator R. A. Barker.
        (Published by the Academia de Marinha, Lisbon,
        1996, as a facsimile, transcript and translation; and
        including the 1965 commentary by Dr João da
        Gama Pimentel Barata, revised and updated from
        his notes on a draft translation in 1986; together
        with further annotation by the translator). <https://
        academia.marinha.pt/pt/edicoes/Paginas/Reedições.
        aspx>. Accessed: 30 November 2022.

Loewen, Brad
1998   The Morticed Frames of XVIth Century Atlantic
        Ships and the «madeiras Da Conta» of Renaissance
        Texts. *Archaeonautica* 14(1):213–222.

Loureiro, Vanessa
2012   Signatures architecturales vs. Spécificités régionales
        au sein de la tradition de construction navale ibéro-
        atlantique »(Architectural Signatures vs. Regional
        Specificities within the Ibero-Atlantic Shipbuilding
        Tradition). e-Phaïstos. *Revue d'histoire des techniques
        (Journal of the History of Technology)* I (1):27–38.

2016   Épaves et espace culturel : L'exemple de l'épave
        du xvie siècle Arade 1 (Algarve, Portugal) et des
        traditions architecturales Ibériques: une première
        approche (Wrecks and Cultural Space: The Example
        of the 16th-century Wreck Arade 1 (Algarve,
        Portugal) and the Iberian architectural Traditions: A
        First Approach). In *Objets et symboles : De la culture
        matérielle à l'espace culturel*. Laurent Dhennequin,
        Guillaume Gernez, and Jessica Giraud, editors, pp.
        115–130. Éditions de la Sorbonne, Paris.

Malcolm, Corey
2001   Lead Hull-Sheathing of the 1622 Galleon *Santa
        Margarita*. *The Navigator: Newsletter of the Mel
        Fisher Maritime Heritage Society* 16(1):3.

Oertling, Thomas
2001   The Concept of the Atlantic Vessel. In Tr*abalhos
        de Arqueologia, 18-Proceedings-International
        Symposium on Archaeology of Medieval and
        Modern Ships of Iberian-Atlantic Tradition: Hull
        Remains, Manuscripts, and Ethnographic Sources: A
        Comparative Approach*, 233–240. Lisbon: IPA.

2007   The Highborn Cay wreck: The 1986 field season.
        The International Journal of Nautical Archaeology
        18(3):244–253.

Ridella, Renato Gianni, Ruth Brown, Marco
Milanese, and Kay Smith.
2018   The *San Juan/Parissona Grossa* - 1581. The
        Identification of a Wreck Found off Sciacca, Sicily,
        through Archaeology and Archives. *Journal of the
        Ordnance Society* 25:36–66.

· · · · · · · · · · · · · · ·

Marine Jaouen
Department of Underwater
Archaeological Research (DRASSM)
French Ministry of Culture
147 Plage de l'Estaque,
13016 Marseille, France

Sébastien Berthaut-Clarac
Research Center on Societies and
Environments in the Mediterranean (CRESEM),
Perpignan University - Via Domitia
52 Avenue Paul Alduy 66860
Perpignan Cedex 9, France

The Bay of Gorée in the Structuring of the Maritime History of the Peninsula of Cape Verde, Africa

Madick Gueye

The cultural and economic interactions during the Atlantic period have engendered manifestations in all regions of the Atlantic Basin. The present is the result of this recent past. In Senegambia, the presence of slave ships in the Bay of Gorée led to a reorientation of trade routes towards the Atlantic. The peninsula of Cape Verde, Africa, thus became an integral part of the Atlantic system where merchants from various horizons converged for the acquisition, production, and consumption of trade goods. The incorporation of this coastal region into the Atlantic processes reshaped the local cultural landscapes. It transformed the pre-existing political system and created a profound sense of identity.

As interações culturais e económicas durante o período atlântico geraram manifestações em todas as regiões da bacia atlântica. O presente é o resultado deste passado recente. Na Senegâmbia, a presença de navios negreiros na baía de Gorée levou a uma reorientação das rotas comerciais em direção ao Atlântico. A península de Cabo Verde, em África, tornou-se assim parte integrante do sistema atlântico para onde convergiam mercadores de vários horizontes para a aquisição, produção e consumo de bens comerciais. A incorporação desta região costeira nos processos atlânticos alterou as paisagens culturais locais. Transformou o sistema político pré-existente e criou um profundo sentido de identidade.

Les interactions culturelles et économiques au cours de la période de l'Atlantique ont engendré des manifestations dans toutes les régions du bassin de l'Atlantique. Le présent est le résultat de ce passé récent. En Sénégambie, la présence de navires de traite des esclaves dans la baie de Gorée a conduit à une réorientation des routes commerciales vers l'Atlantique. La presqu'île du Cap-Vert, en Afrique, est ainsi devenue une partie intégrante du système atlantique où les marchands de divers horizons ont convergé pour l'acquisition, la production et la consommation de biens de commerce. L'intégration de cette région côtière dans les processus de l'Atlantique a remodelé les paysages culturels locaux. Il a transformé le système politique préexistant et créé un profond sentiment d'identité.

Introduction

This paper examines the role of the Bay of Gorée in the structuring of the economic and political history of the peninsula of Cape Verde, Africa. Since the 15th century, the Lebou's region has become an essential link in the maritime transactions linking Senegambia and the rest of the world. More than just a network for the movement of goods and people, backed by commercial practices, the exploitation of the Bay of Goree by the Atlantic transoceanic system has shaped new physical and cultural landscapes, with new and very specific identities in the peninsula of Cape Verde. This area served as a point of economic and cultural interaction between sea and land, between the coast and the hinterland, and between Senegambia and the rest of the world. For more than 400 years, interactions between Africans and Europeans in this region took place via slave ships anchored in this bay. This paper is an analysis of the maritime cultural landscape of the region during the Atlantic trade. It seeks to highlight the mechanism of transatlantic interactions in the region and the evolution of economic and political landscapes during this key moment in which the contours of nascent capitalism were being shaped.

The Bay of Gorée: A Decisive Element in the Commercial Dynamics of the peninsula of Cape Verde

The maritime cultural landscape here is defined as the space encompassing both the Bay of Gorée and the adjacent coastline, integrating the interactions between African traders and European navigators in the village of Begne, with each other and with the sea, and the various cultural processes that have affected the region over the course of the long historical trajectory (Westerdahl 1991, 1992, 1994). The Bay of Gorée (Figure 1) was one of the focal points of this traffic, linking the coastal hinterland (Cayor) and the Atlantic coast (the peninsula of Cape

Verde) on the one hand, and Senegambia and the rest of the world, notably Europe and North America, on the other. It thus constitutes a special space for the Lebou community, which defines it as its partial structure of reference supporting its economic activity during the Atlantic trade. Landscape archaeology provides an interpretive framework for analyzing the question of how the peninsula of Cape Verde and its history lie in the bay of Gorée. It presents an efficient method of integrating the physical and social environments in which the Atlantic trade operated in the region.

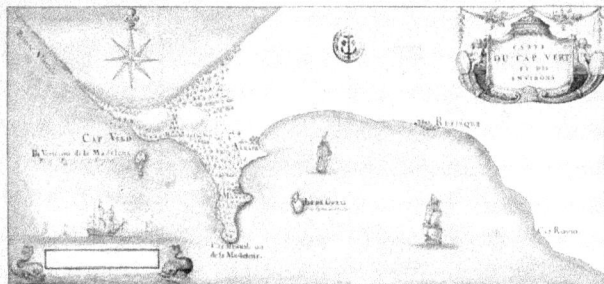

FIGURE 1: The Bay of Gorée: Meeting Point between Europe and the presque'île du Cap-Vert. (Stanislas de Boufflers 1787; Figure courtesy of the Bibliothèque nationale de France [National Library of France.])

This space of interaction represents a social construction that has removed the cultural and physical boundaries between Europe and the region, and participated in the restoration of the mental geographies of space of local communities (Ogundiran and Falola 2010:41). Every aspect of the maritime landscape of the Lebou's region, be it oceanographic; climatic and geomorphological phenomena; the location of villages, such as Bègne and Dakar; the establishment of the post of the Goree trading post; the anchorage sites; the aiguade (water needle) of Hann; the archaeological remains located on the sites of shipwrecks (even the most insignificant ones); the people who interacted in the Bay of Hann and at the level of the village of Begne; and the activities that animated the exchanges, constitutes an essential component in the study of the maritime past of this region.

The Bay of Gorée, combined with historical and archaeological data, provides insight into the long-term patterns of human behavior that punctuated the peninsula of Cape Verde during transatlantic trade. These are the evolving historical processes within the Lebou's region and with their maritime environment. Understanding how the region interacted with the bay is an effective way to analyze the socio-political transformations that occurred in the region during the Atlantic

period and the foundations of modernity (Crumley and Marquardt 1990; Thiaw and Mack 2020). This article approaches maritime landscapes (Westerdahl 1994) as a theoretical framework for explaining the evolution of cultural landscapes in the peninsula of Cape Verde during transatlantic interactions (Richard 2012).

Perceived as simply a physical setting, the Bay of Gorée is defined as a space over which established traders in the peninsula of Cape Verde moved and acted with trading ships from Europe. It is socially produced and made significant by the traders who interacted there and the economic and cultural activities that took place. The commercial dynamism and changing political landscapes of the region during the Atlantic period is relative to the opportunities offered by its maritime façade.

The Peninsula of Cape Verde: an Integral Part of Atlantic Processes

The history of the peninsula of Cape Verde during the Atlantic Period can be summed up essentially in commercial practices, maritime navigation, cultural interactions, and identity recompositions (Gueye 2023; Thiaw 2008), because the region was the place of convergence for many commercial actors in Senegambia. Due to the increased presence of European ships in the bay and the commercial dynamics of the trading post of Gorée, which polarized all the trading points of the Petite-Côte, Begne ended up capitalizing on a good part of the maritime commercial activities of Cayor. The space of exchange and interaction that is the Bay of Gorée, as well as the village of Begne and the marigot of Hann on the coast, and all the activities that took place there, are defined as the maritime cultural landscape of the peninsula of Cape Verde.

The Bay of Gorée played a fundamental role in transatlantic interactions and in structuring the maritime history of the peninsula of Cape Verde. For more than four centuries, trade between Europeans and the Lebou's region was carried out via slave ships anchored near the shore. As such, it served for 500 years as a point of interaction between European traders and Africans from various parts of Senegambia, sea and land, and is the scene of Atlantic commercial and cultural encounters between the region and Europe (Horlings 2011). In the peninsula of Cape Verde, these interactions represent a complex dialectic composed of visible and invisible actions, discourses, and movements, the birth of which is explained by social differences and cultural particularities (Richard 2015). Far from interpreting the Atlantic side of the peninsula as a structural and political history (Barry 1998), this paper conducts an anthropological interpretation of the maritime history of the Dakar by making an analysis of how the processes of the Atlantic are transposed into the reality of everyday life by re-examining how practices and ideas are reshaped in the Lebou's community.

The fundamental reconfigurations observed in the socio-political structures of the peninsula of Cape Verde are the obvious consequences of the arrival of slave ships in the Bay of Gorée and the interactions that followed. The study of commercial landscapes during the Atlantic shows that the region did not remain passive in the face of the demands of the nascent capitalist economy. The transatlantic economic activities of Europe had established very strong connections with the political and economic powers of Cayor (Becker and Martin 1975; Thilmans and de Moraes 1976), in particular those of the peninsula of Cape Verde. European navigators were required to offer gifts to local chiefs, notably the Alcati resident in Begne, and to pay taxes for anchoring their ships in the Bay of Gorée and other taxes for refreshments offered by the marigot of Hann and for trade (de Moraes 1993) (Figure 2 and 3).

FIGURE 2: L'Alquier collecting taxes in Begne. (Figure courtesy of the Bibliothèque nationale de France [National Library of France.])

Indeed, the relationships (intra and extra) of the area are strongly reshaped by Atlantic processes. Due to the nature of the Bay of Gorée and its exploitation, and especially the role of the island of Gorée as a commercial warehouse, the region has a growing involvement in the establishment and unfolding of the world economy. It has, thus, undergone profound social and spatial transformations due to the changing economic and political landscapes over the course of a long historical trajectory (Gueye 2023).

FIGURE 3: Scene of capture of slaves in Begne (Villeneuve, René Geoffroy de (RGV), 1767–1831. L'Afrique, ou histoire, mœurs, usages et coutumes des africains. Paris : Nepveu, no. 26, 1814 [Part of Villeneuve, René Geoffroy de (RGV), 1767–1831. Africa, or history, manners, uses and customs of Africans. Paris: Nepveu, library, passage des Panoramas, no. 26, 1814]). (Figure courtesy of the Library Company of Phildelphia.)

Begne: An Ancient Commercial Town "Discovered"

The presence of trading ships in the Bay of Gorée led to changes in Cayor trade patterns. Once oriented toward the trans-Saharan routes, part of the Cayorian trade networks were now focused on Begne, a small, isolated settlement on the periphery of the Saharan trade system, which became an important commercial center for the purchase and distribution of trade products (Gueye 2023). The fishermen and farmers of the peninsula, as well as others from the interior, became experienced traders and middlemen who benefited from trade at the *aiguade de Hann* (de Moraes 1993:54). In fact, Drick Ruiters describes the commercial activities in this town as follows in 1602:

"... the people who live there go, with what they have to sell, to the city, which is located inside the Cape, facing the island of Goree, which the ships anchor in the roadstead. Nothing in particular is purchased there except for iron skins and brandy, which is the most important merchandise. Yellow serge and an assortment of very wide and flat

basins, in which a man can wash himself, are also sold there. You can also trade red serge...Silesian cloth, amber and coral beads, nurembergeois ware (small wrought metal objects), glass beads. For beads, you will get only a few palm wine, and with difficulty, but there is an assortment of eight-sided crystal beads, thanks to which it will be easy to get hens and other refreshments. This people is so fond of brandy that it regrets that the bottles of brandy do not always hang around its neck" (Ruiters 1602, as quoted in de Moraes 1993:70).

The peninsula of Cape Verde became an important supplier of trade products and refreshments, and constituted a society of consumption of European goods (Thiaw 2000). Responding to the demands of mercantile capitalism, Begne contributed in its own style and in a decisive way to the development of the modern world system and the mosaic of lives and cultures that emerged (Stahl 1999). The trade between the European navigators and the traders established in the region settled and developed rapidly in Begne in relative stability. The resources offered by the *aiguade de Hann*, and the nature of the shallow and calm bay, made the area the main anchorage for trading ships. Using local canoes, European traders would travel to the coastline to pay royalties and offer gifts to the Alquier to trade. The local expertise in maritime navigation made possible the Atlantic commercial interactions in the peninsula of Cape Verde. Begne suddenly became the gateway to the Goree trading post, a very dynamic commercial city for the acquisition and distribution of trade products.

The Appearance of a Territorial Attachment and a Manifest Affirmation of Identity

The development of transatlantic trade in the peninsula of Cape Verde has led to what Guyer (1993) calls the logic of decomposition. The purchase, production, and consumption of more and more trade products created and reinforced a local power that was out of phase with that embodied by the central state of Cayor. A series of rebellions and repressive attacks by the Damel followed. Indeed, the development of Atlantic trade in the city of Begne ended up installing a feeling of independence. This situation corresponded to the rise in power of Cayor, which derived its wealth from the exploitation of peasants and commercial exchanges at the level of the trading posts of Rufisque and Begne (Becker and Martin 1975). A clash of ambitions collided with the relations

between the central power and this small coastal community. The Cayor thus entered into a spiral of social disintegration, raids and political crises.

The peninsula of Cape Verde was the architect of one of the greatest changes that reshaped the political geography of the Cayor kingdom. Directly adjacent to the Atlantic coastline, and an integral part of maritime exchanges through the Bay of Gorée, the Lebou region evolved between two political systems during the Atlantic exchanges. From small communities of villages governed by the lamanal system from the 15th century onwards and under the tutelage of Cayor, the region finally gained its independence in 1790 through cannonades after several decades of boycotts and wars against the central state.

Conclusion

The analysis of the maritime cultural landscape of the peninsula of Cape Verde during the Atlantic trade period provides a general overview of the maritime history of the region. Concomitant with the lesser-known Indigenous social, political, and economic transformations in the region, this study provides elements of the overall context in which the Bay of Gorée was exploited by the various forces governing nascent capitalism. The entry of the peninsula into the world economy through the opportunities offered by the Bay of Gorée and the rise of Islam in the region remain the facts that strongly shook Cayor and led to the restructuring of its political map.

References

BARRY, BOUBACAR
1998   *Senegambia and the Atlantic Slave Trade.* Cambridge University Press. NY.

BECKER, CHARLES, AND VICTOR MARTIN
1975   Kayor et Baol: Royaumes sénégalais et traite des esclaves au XVIIIe siècle (Kayor and Baol: Senegalese Kingdoms and the 18th century Slave Trade). *Revue française d'Histoire d'Outre-Mer (French Review of Overseas History)* 62:270–300.

CRUMLEY, CAROLE L., AND WILLIAM H. MARQUARDT
1990   Landscape: A Unifying Concept in Regional Analysis. In *Interpeting Space: GIS and Archaeology,* edited by Kathleen M. S. Allen, Stanton W. Green and Ezra B. W. Zubrow, pp. 73–79. Taylor & Francis Publishing: London, UK.

DE MORAES, NAZE I.
1993   À la découverte de la petite Côte au XVII (Sénégal et Gambie) [Discovering the Petite-Côte in XVII (Senegal and Gambia)]. Tome I: 1600–1621. University Cheikh Anta Diop-Dakar-IFAN, Dakar, Senegal (Cheikh Anta Diop University, Fundamental Institute of Black Africa, Dakar, Senegal).

GUEYE, MADICK
2023   L'étude du paysage culturel maritime de la presqu'île du Cap-Vert pendant la traite atlantique (Investigating the Maritime Cultural Landscape of the Presqu'île of Cap-Vert During the Atlantic Slave Trade). Thèse de Doctorat (PhD), Aix-Marseille University, Marseille, France.

GUYER, JANE I.
1993   Wealth in People and Self-Realization in Equatorial Africa. Man 28(2):243–265.

HORLINGS, RACHEL
2011   Of His Bones are Coral Made: Submerged Cultural Resources, Site Formation Processes, and Multiple Scales of Interpretation in Coastal Ghana. Ph.D. Dissertation, Department of Anthropology, Syracuse University.

OGUNDIRAN, AKINWUMI, AND TOYIN FALOLA
2010   Pathways in the Archaeology of Transatlantic Africa. In Archaeology of Atlantic Africa and the African Diaspora, edited by Akinwumi Ogundiran and Toyin Falola, pp.3–45. Indiana University Press, Bloomington, IN.

RICHARD, FRANÇOIS G.
2012   Political Transformations and Cultural Landscapes in Senegambia during the Atlantic Era: An Alternative View from the Siin (Senegal). In Power and Landscape in Atlantic West Africa, Archaeological Perspectives, edited by J. Cameron Monroe and Akinwumi Ogundiran, pp. 78–114. Cambridge University Press, Cambridge, England.

2015   The African State in Theory: Thoughts on Political Landscapes and the Limits of Rule in Atlantic Senegal (and Elsewhere). In Theory in Africa, Africa in Theory: Locating Meaning in Archaeology, edited by Stephanie Wynne-Jones and Jeffrey B. Fleisher, pp. 201–231. Routledge Publishing, Abingdon, England.

STAHL, ANN B.
1999   The Archaeology of Global Encounters Viewed from Banda, Ghana. African Archaeological Review 16(1):5–81.

THIAW IBRAHIMA ET DEBORAH MACK L.
2020   Atlantic Slavery and the Making of the Modern World: Experiences, Representations, and Legacies. Current Anthropology 61(22).

THIAW, IBRAHIMA
2000   Impact of the Black Trade in the Upper Senegal River: Archaeology of Afro-European Interactions in the Gajaaga and Buundu in the 18th and 19th centuries. In Saint-Louis and the Slavery. Edited by D. Samb (Ed.). Cheikh Anta Diop University, Fundamental Institute of Black Africa, Dakar, Senegal.

2008   Every House has a Story. In Africa, Brazil and the Construction of Trans-Atlantic Black Identities. Africa World Press, Trenton, NJ.

THILMANS, G., AND NIZE I. DE MORAES
1976   Villault de Bellefond sur la côte occidentale d'Afrique. Les deux premières campagnes de l'Europe (Villault de Bellefond on the West Coast of Africa. The First Two Campaigns in Europe) 1666–1671. Bulletin de l'IFAN, B, 38(2):257–299.

WESTERDAHL, CHRISTER
1991   Norrlandsleden: The Maritime Cultural Landscape of the Norrland Sailing Route. In Aspects of Maritime Scandinavia AD 200–1200. Proceedings of the Nordic Seminar on Maritime Aspects of Archaeology. Edited by O. Crumlin Pedersen, Wiking Ship Museum, Roskilde, Denmark.

1992   The Maritime Cultural Landscape. The International Journal of Nautical Archaeology 21(1):5–14.

1994   Maritime Cultures and Ship Types. The International Journal of Nautical Archaeology 23(4):265–270.

· · · · · · · · · · · · · · ·

Madick Gueye
Laboratoire d'Archéologie Médiévale
et Moderne en Méditeranée
Aix-Marseille Université58 Boulevard Charles Livon
13100 Aix-en-Provence
France

Wood Analysis from the IDM-013 Shipwreck

Stéphanie Wicha, David L. Conlin, Marc-André Bernier

In 2020, a team of archaeologists from the Slave Wrecks Project recovered several wood samples from the wreck IDM-013 in Mozambique. This article discusses the scientific findings and implications for the possible origin of the wreck, based on the analysis of the wood used for various elements of the hull, including the ceiling planks and the frames. This ship is presumed to be the French slave ship, L'Aurore, built in Saint-Malo shipyards in 1783, and sunk off the coast of Mozambique in 1790, with a cargo of 600 enslaved people. Identification of the species of wood and of their biogeographic distribution aims to verify this French provenance hypothesis. The authors also evaluated the dendrological potential of these remains and discuss implications for future research.

Em 2020, uma equipa de arqueólogos do Slave Wrecks Project recuperou várias amostras de madeira do naufrágio IDM-013 em Moçambique. Este artigo discute os resultados científicos e as implicações para a possível origem do navio, com base na análise das madeiras utilizadas em vários elementos do casco, incluindo as tábuas forro interior e as balizas. Presume-se que este naufrágio seja o navio negreiro francês L'Aurore, construído nos estaleiros de Saint-Malo em 1783, e afundado ao largo de Moçambique em 1790, com uma carga de 600 pessoas escravizadas. A identificação das espécies de madeira e da sua distribuição biogeográfica tem como objetivo verificar esta hipótese de proveniência francesa. Os autores avaliam também o potencial dendrológico destes vestígios e discutem as implicações para a investigação futura.

En 2020, une équipe d'archéologues du Slave Wrecks Project a récupéré plusieurs échantillons de bois de l'épave IDM-013 au Mozambique. Cet article traite des découvertes scientifiques et des implications pour l'origine possible de l'épave, basée sur l'analyse des bois utilisés dans divers éléments de la coque, y compris le vaigrage et les membrures. Ce navire est présumé être le navire de traite d'esclaves français, L'Aurore, construit dans les chantiers navals de Saint-Malo en 1783, et coulé au large des côtes du Mozambique en 1790, avec une cargaison de 600 esclaves. L'identification des essences de bois et de leur distribution biogéographique vise à vérifier cette hypothèse de provenance française. Les auteurs ont également évalué le potentiel dendrologique de ces restes et discutent des implications pour la recherche future.

Introduction

This paper presents the first archaeobotanical results conducted on a shipwreck discovered off the coast of Mozambique, on the eastern coast of the African continent. This wreck is probably the remains of the enslaved trade ship *L'Aurore*, which was built in 1783, in Saint-Malo in the northwestern part of France. In the 18th century, the port of Saint-Malo was renowned for its merchant vessels and privateers, but it found a secondary activity in the slave trade (Mettas 1984).

L'Aurore sank off the coast of Mozambique in February of 1790, with a cargo of 600 enslaved, 360 of which drowned in the wrecking event. The excavation of this wreck is being conducted through the Slave Wrecks Project, a research program studying the slave trade through shipwrecks, (National Park Service Submerged Resources Center). The study of this shipwreck is in its early stages. The discovered remains are poorly preserved ,with some sections of the wreck having been uncovered, exhibiting elements of the hull with iron nails and lead caulking. In 2020, six samples of wood were taken and four in 2022. It should be noted that the first samples received for archaeobotanical analysis were particularly fragmented, which did not favor a dendrochronological study. Nine wood samples were collected from various parts of the ship for analysis and one from the remains of a barrel stave. These samples were stored in plastic bags with water and kept in the refrigerator.

It is hoped that the identification of the species of timber used in the construction of the ship will make it possible to confirm or deny its French provenance. Indeed, the questions related to identification of the wood can provide answers to the questions: Which taxa were used and where did they come from? The hypotheses concerning the possible construction areas of the wrecks are based on the native distribution of the different tree species used in their construction. The potential for dendrochronological dating of these wood samples was also evaluated. Indeed, the dating of the wood samples should also allow researchers to specify the place of construction of the ship, a process called dendro-provenance as will be discussed later.

Methods

Nine samples were taken from the wreck, originating from different structural elements of the hull: frames, ceiling planking, outer planking, and inner strake, or stringer. The wood remains are, for the most part, fragmented, badly preserved, and eroded (Figure 1). The sample for the identification of tree species was done manually with a razor blade. Thin sections in three planes (transverse, tangential, and radial) were cut and each sample was mounted on slides. Identification of different wood species was possible with anatomical criteria by observing the transverse plane under a microscope to examine vertical elements for support and storage of sap. This first observation provides a quick determination of homogeneous wood (softwood) from heterogeneous wood (hardwood). Analyses were performed using light microscopy and an anatomical atlas of European trees (Schweingruber 1978, 1990).

Results

The wood analyses revealed the use of four taxa: two types of softwood, Norway spruce/larch (*Picea abies* Karst./*Larix decidua* Mill.) and maritime pine (*Pinus pinaster* Aiton) for ceiling and outer-hull planks; and two species of hardwoods, cork oak/green oak (*Quercus*

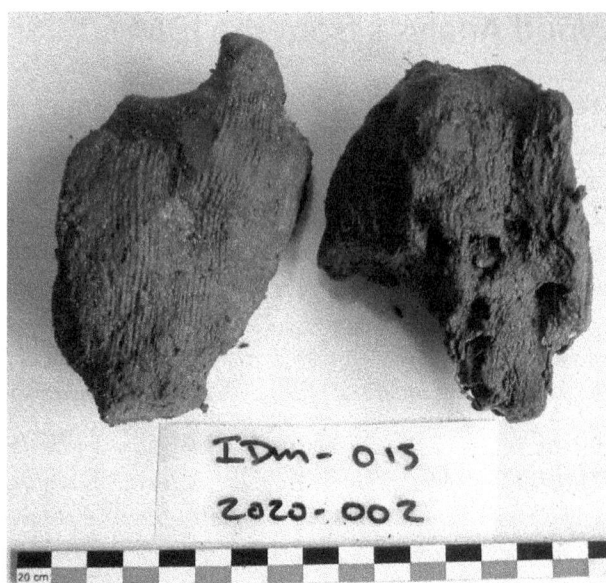

FIGURE 1. Example of the poor preservation of wood on the wreck. (Photo by the author, 2021.)

suber L./*Quercus Ilex* L.), and deciduous oak for the framing (Table 1). The wood species from spruce/larch and cork oak/green oak are anatomically indistinguishable. It should be noted that the number of wood species used is remarkable, with nine structural elements and four taxa identified, which is high and reflects a heterogeneous construction. Regarding the implementation of these species in the ship's frame structure, enough samples to have a global view of its construction were not taken. However, the wood species present are used wisely. Oak, for example, was favored by the Navy for

LABEL	LOCATION	DESCRIPTION	WOOD SPECIES	TYPE OF CUT	NUMBER OF RINGS
IDM-013-2020-001	Trench 1, A-13, A-14	Sample of dark hard wood (oak??) that is connected together with iron fasteners.	cork oak/green oak	?	?
IDM-013-2020-002	Trench 1, A-13, A-14	Sample of light soft wood (pine?) that is not connected and sits next to area where sample 001 was collected.	maritime pine	quartered	approx. 68
IDM-013-2020-003	Trench 2, A-18, B-16	Sample of barrel stave	deciduous oak	Halved	approx..40
IDM-013-2020-007	Trench 3, A34, A27	Sample of interior strake or stringer	maritime pine	?	approx.13
IDM-013-2020-008	Trench 3, A34, A27	Sample of frame timber	deciduous oak	?	approx.10
IDM-013-2020-009	Trench 3, A34, A27	Sample of possible outer hull timber	maritime pine	?	approx.10
IDM-013-2022-013		Sample of ceiling	spruce/larch	the heart of the wo	6 tree rings
IDM-013-2022-013		Sample of plank hull	spruce/larch	branch start	approx.13 tree rings
IDM-013-2022-013		Sample of frame timber	cork oak/green oak	quartered	approx. 20 tree rings

TABLE 1. Synoptic table of wood samples.

the manufacture of the frame of the ship. Green oak, unlike cork oak, is considered one of the best woods by the Navy, but its small size, it is only used in smaller hull elements. These hardwoods are durable and resistant to both heavy stress and humidity, and their natural curves are suitable for shaping the frames. As for softwoods, they offer straight trunks over long lengths which are flexible and suitable for planking. In particular, spruce is the softwood that has the best ratio of mechanical qualities to lightness (Lieutaghi 2004:547). Moreover, it resists shocks well and its trunk is rectilinear.

Discussion About the Origins of the Boat

The identified species (maritime pine, spruce, and oak) are available in France, both on the Mediterranean and Atlantic coasts. The documentation regarding a slave ship called *L'Aurore* was revealed in the archives of the port of Rochefort (Boudriot 1984) in the work of shipbuilder, Hubert Penevert, who spent many years in this city. From these documents came the principal data for Boudriot's reconstruction. More recent archival research following on the footsteps of Mettas (1984) showed that *L'Aurore* was lost in Senegal and built in Saint-Malo in 1783 (SHD de Brest:1 P7-14). It is reasonable to believe this ship was built in Saint-Malo. The taxa identified are, for the most part, present in the environment of this port, maritime pine, cork oak, and oak (Visset and Bernard 2006:11). However, spruce was undoubtedly imported, probably from the Massif Central forests via the hydrological network (Miras et al. 2010; De Beaulieu and Reille 1992; De Beaulieu et

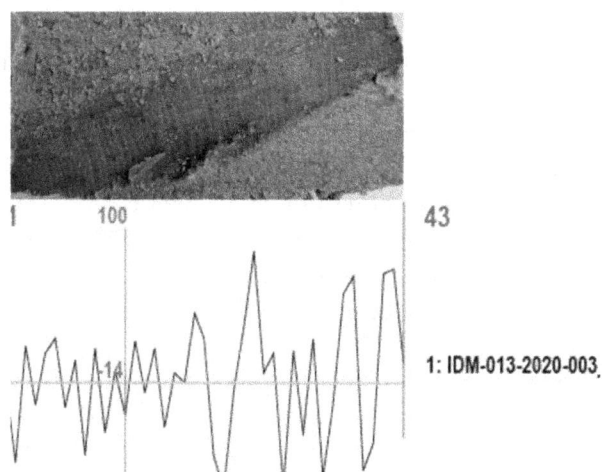

FIGURE 2. Construction of maritime pine individual curve. The numbers above the vertical straight lines correspond to the relative ring numbers (Photo by the author, 2021).

al. 1993). If the ship discovered off Mozambique was constructed in Saint-Malo, it is not possible to exclude either Atlantic or Mediterranean provenances (French or Italian).

Dating and Origin

To specify this provenance, dating using dendrochronology was attempted. The guidelines for this dating are developed according to climatic regions, and when dating a piece of timber to circumscribe the supply forests is possible, discussion of the dendro-provenance as alluded to previously is possible (Lambert 2014). However, for this type of dating, many issues, including the insufficient quantity of samples (9) for four different taxa, were encountered; moreover, these timbers have a positive growth and the ring series are too short (between 6 and 40 tree rings), which makes them unsuitable for dendrochronological dating. These large rings reflect growth in an environment that is adequate for the tree, with good soil and regular watering. This can be an indication of provenance, particularly for the cork oak/green oak, which in the Mediterranean is submitted to a more restrictive system that has a real impact on its growth. In fact, it is important to note that for hardwoods, a large growth increases wood quality.

Only one sample of maritime pine wood presented enough tree rings (68) for an attempt at dating, but without any convincing results (Figure 2). Indeed, this individual series is relatively short, which means that the climatic signatures are not very characteristic, and we do not have a specific reference for this species of wood, which reduces the statistical result. Thus, several dates appear that it is impossible for us to discriminate from each other. It seems, therefore, necessary to multiply the samples to obtain a significant and long average for each species to ensure a reliable dating.

As dating these woods by dendrochronology was not possible, ^{14}C dating was initiated. These dates were graciously provided by Christine Hatté from the Laboratory of Climate and Environment Sciences (UMR CEA CNRS UVSQ 8212). For these analyses we worked on three different pieces of framework: a frame, an outer plank and a ceiling plank. On each of these pieces we took two samples, one close to the heart (the oldest part) and another corresponding to the last tree ring preserved on the piece (thus the most recent). It is a method by which radiocarbon dates are compared to the 'wiggles' of the calibration curve (Hatté et al. 2023; Bronk Ramsey et al. 2001). Radiocarbon dendrochronology wiggle-match models make use of the known sequence of ^{14}C-dated sub-samples of wood and the known chronological interval between them to better refine the date of an archaeological event, e.g., the felling of a tree. The results of this analysis gives a date in the middle of the 17th century, which is not consistent with the date of construction of *L'Aurore* at the end of the 18th century. At best, there is a gap between these two dates (1642 and 1783) of 141 years, which is significant. However, several reasons can explain this discrepancy. The method of debit of the pieces, namely that for the planks the debit can be in the heart of the tree, in this case the trunk is squared and the last rings of the tree are lost. The alteration of the pieces of framework that were observed, the frames are eroded, and the last rings are not present. Note that for the oldest trees, there is an effect of senescence, which is to say that the tree in aging forms smaller rings. This is why when the periphery of the trunk is missing, many rings can be missing. Finally, at the time of the construction of the ship (1783), the stocks of wood for shipbuilding existed and sometimes between the felling of the tree, the transport of the wood to the dockyard and the construction of the ship, many years can pass. This is why, if these ^{14}C dating results are surprising in relation to the year of construction of *L'Aurore*, they do not allow the exclusion of the hypothesis of the discovery of this boat.

Final Thoughts

At present, the identification of the timbers from the wreck discovered off the coast of Mozambique do not contest the assumption put forward by archaeologists that it is the merchant ship *L'Aurore*. However, this is not definitive. Indeed, the nine structural components that were analyzed were made of wood species (oak and pine) found in the nearby area of the ship's building site, Saint-Malo. With two exceptions, the elements are made of spruce, a species of wood imported and available in the French region of the Massif Central. The ship, however, could just as easily have come from a port on the French Mediterranean coast (Marseille, Sète), where all the taxa discovered are available, and in addition are also found in Italy. In the 18th century, the spruce (or larch) represents a tree species with a more limited biogeography, populating the forests of the Alps, and is not endemic in the Pyrenees, recently introduced in the framework of the reforestation.

This problem related to the biogeography of the identified taxa, and the impossibility to carry out dendrochronological dating on the wood samples, prevents

a confirmed construction site for the presumed *L'Aurore*. Therefore, even if the construction of this wreck can possibly be at the port of Saint-Malo, more wood sample analysis is required. Furthermore, to answer the question of where the ship was built, additional attempts were made to find out more about the use of wood species for the construction of these merchant ships. A monograph by J. Boudriot on *L'Aurore* (1984:28) attempts to describe and evaluate the lumber used for a comparable merchant ship c. Boudriot provides a description of the quantity of wood required, the type of wood used, and its usage in the structure (frame, hull, mast, etc.). It seems that oak is used for both the hull planking and the frames of the ship, pine or fir being used for the masts. In this respect, discoveries discussed here diverge, on the one hand by the variety of taxa identified and on the other hand, using spruce for the hull planking rather than oak. This questions the economic value of this type of ship, a merchant vessel, and its construction being understandably secondary to the construction of warships; it may be the case that it benefits economically from wood of lesser quality. For example, there is a large variety of wood species used, with four taxa amongst the 10 samples of frame components studied. However, this was not a merchant ship like the others, because it was destined for the slave trade and was able to carry a cargo of 600 men, women, and children. Undoubtedly, greater financial benefit existed for the shipowner that this cargo reach its destination. These implied constraints in the construction of the ship were to reduce the cost of each crossing and limit the mortality of the enslaved people (Boudriot 1984:18). For these reasons, the timber for such a ship had to have particular characteristics and specifically be lighter, which could explain the use of spruce for the hull planking.

To date, dendrochronological dating has not been possible for several reasons. These include the insufficient number of samples in relation to the poor conservation of timber and the large number of species with the low number of growth rings (less than 50). The ^{14}C dating results are far from the probable date of the construction of the ship *L'Aurore* and do not allow the authors to confirm or deny that the shipwreck discovered off the coast of Mozambique is really this ship. In view of the interest of such research, it is important to carry out a systematic sampling of the structural parts of this vessel An increased number of samples will help answer the various questions of dating and provenance, and also technical and economic questions specific to this type of merchant ship transporting the enslaved. This work allows research into the timber of hulls and to identify merchant ships that until now were little studied.

References

BRONK RAMSEY, CHRISTOPHER; JOHANNES VAN DER PLICHT, AND BERNHARD WENINGER
2001   "Wiggle Matching" Radiocarbon Dates. *Radiocarbon* 43(2A): 381–389.

BOUDRIOT, JEAN
1984   *Traité et navire négrier: l'Aurore, navire de 280 tx, 1784. (Treaty and Slave Ship: the Aurore, Ship of 280 tx, 1784).* Boudriot: Paris, FR.

DE BEAULIEU, JEAN-LOUIS AND REILLE, MAURICE
1992   Long Pleistocene Pollen Sequences from the Velay Plateau (Massif Central, France). *Vegetation History and Archaeobotany* 1:233–242.

DE BEAULIEU, JEAN-LOUIS, KOSTENZER, JOHANNES AND REICH, KIM
1993   Dynamique forestière dans la haute vallée de l'Arve (Haute Savoie) et migrations de Abies et Picea dans les Alpes occidentales (Forest Dynamics in the Upper Arve valley (Haute Savoie) and Migrations of Abies and Picea in the Western Alps). *Dissertationes Batanicae* 196:387–398.

HATTÉ, CHRISTINE, MAURICE ARNOLD, ARNAUD DAPOIGNY, VALÉRIE DAUX, GEORGETTE DELIBRIAS, DIANE DU BOISGUEHENEUC, MICHAEL FONTUGNE, CAROLINE GAUTHIER, MARIE-THÉRÈSE GUILLIER, JÉRÉMY JACOB, MICHAEL JAUDON, EVELYNE KALTNECKER, JACQUES LABEYRIE, CLAUDÉE NOURY, MARTINE PATERNE, MONIQUE PIERRE, BRIAN PHOUYBANDHYT, JEAN-JACQUES POUPEAU, JEAN-FRANÇOIS TANNAU, FRANÇOIS THIL, , NADINE TISNÉRAT-LABORDE, AND HÉLÈNE VALLADAS
2023   Radiocarbon Dating on ECHoMICADAS, LSCE, Gif-sur-Yvette, France: New and Updated Chemical Procedures. *Radiocarbon*.

LAMBERT, GEORGES
2014   *Dendro-assistance à la détermination de la provenance géographique des bois archéologiques (Dendro-Assistance in Determining the Geographical Origin of Archaeological Woods).* ARCADE Approche diachronique et Regards croisés: Archéologie, Dendrochronologie et Environnement, Aix-en-Provence, France. pp.33–49.

LIEUTAGHI, PIERRE.
2004   *Le livre des arbres, arbustes & arbrisseaux. (The Book of Trees, Bushes, and Shrubs).* Actes Sud: Arles, FR.

METTAS, JEAN
1984   *Répertoire des expéditions négrières françaises au XVIIIème siècle (Directory of French Slave Expeditions in the 18th century),* 2 vol. Société d'Histoire Française d'Outre-Mer: Paris, FR.

MIRAS, YANNICK, PASCAL GUENET, AND HERVÉ RICHARD
2010   *La genèse du paysage culturel du plateau de Millevaches (Limousin, Massif central, France): plus de 2000 ans d'histoire révélés par l'analyse pollinique (The Genesis of the Cultural Landscape of the Millevaches Plateau (Limousin, Massif Central, France): More than 2000 Years of History Revealed by Pollen Analysis).* Paysage en Limousin. Presses Universitaires de l'Université de Limoges, Limoges: 99–124.

SCHWEINGRUBER, FRITZ
1978   *Microscopic Wood Anatomy: Structural Variability of Stems and Twigs in Recent and Subfossil Woods from Central Europe.* Zubal-Books: Cleveland, OH.

1990   *Anatomy of European Woods.* Paul Haupt: Bern und Stuttgard, DE.

VISSET, LIONEL AND BERNARD, JACQUES
2006   Évolution du littoral et du paysage, de la presqu'île de Rhuys à la rivière d'Étel (Massif armoricain – France), du Néolithique au Moyen Âge (Evolution of Coastal Landscape, on the Rhuys Peninsula at the River Etel (Massif Armoricain, France), from the Neolithic to the Midddle Ages). *ArcheoSciences* 30(15):143–156.

· · · · · · · · · · · · · · ·

Stéphanie Wicha
Société Archéobois
50 Rue Dragon
13006 Marseille, France

David L. Conlin
National Park Service Submerged Resources Center
12795 W. Alameda Parkway
Lakewood, Colorado 80228

Marc-André Bernier
Manager, Underwater Archaeology Service
Parks Canada (retired)
Government of Canada
1800 Walkley Road, Ottawa
Ontario K1H 8K3

Updated Archaeological Documentation of the Galleon *Santíssimo Sacramento* (1668): Interpretation of Shipwreck Site Formation Processes as an Aid to the Preservation of Underwater Cultural Heritage

Beatriz Bandeira

The following paper presents an analysis undertaken as part of dissertation research into the conservation of the Galleon Santíssimo Sacramento *(1668), with the goal of strengthening underwater cultural heritage in Brazil. The Muckelroy model of shipwreck site formation processes is interpreted through records in loco and videos recorded of the site over a five-year period. The results show both stability and changes from the original documentation in 1978 by Ulisses Pernambucano de Mello Neto. By identifying the processes at work on the site, the case is made that the conservation and protection of underwater cultural heritage is vital to preserve the history and culture of Brazil. While Brazil does not currently recognize the 2001 United Nations Educational, Scientific and Cultural Organization (UNESCO) Convention, considering submerged sites as allies to the preservation of the oceans might be an alternative.*

Este artigo apresenta uma análise realizada no âmbito de uma dissertação sobre a conservação do Galeão Santíssimo Sacramento *(1668), com o objetivo de fortalecer a gestão do património cultural subaquático no Brasil. O modelo de Muckelroy sobre os processos de formação de sítios de naufrágio é aplicado através da análise de registos in loco e vídeos gravados no sítio durante um período de cinco anos. Os resultados mostram estabilidade e mudanças em relação à documentação original de 1978, de Ulisses Pernambucano de Mello Neto. Ao identificar os processos em curso no sítio, defende-se que a conservação e a proteção do património cultural subaquático são vitais para a preservação da história e da cultura do Brasil. Embora o Brasil não reconheça atualmente a Convenção da Organização das Nações Unidas para a Educação, a Ciência e a Cultura (UNESCO) de 2001, considerar os sítios submersos como aliados da preservação dos oceanos pode ser uma alternativa.*

L'article suivant présente une analyse entreprise dans le cadre de la recherche de thèse sur la conservation du galion Santíssimo Sacramento *(1668), dans le but de renforcer le patrimoine culturel subaquatique au Brésil. Le modèle Muckelroy des processus de formation des sites de naufrage est interprété à l'aide de dossiers in loco et de vidéos enregistrées du site sur une période de cinq ans. Les résultats montrent à la fois la stabilité et les changements par rapport à la documentation originale de 1978 par Ulisses Pernambucano de Mello Neto. En identifiant les processus à l'œuvre sur le site, il est fait valoir que la conservation et la protection du patrimoine culturel subaquatique sont vitales pour préserver l'histoire et la culture du Brésil. Bien que le Brésil ne reconnaisse pas actuellement la Convention de 2001 de l'Organisation des Nations Unies pour l'éducation, la science et la culture (UNESCO), considérer les sites submergés comme des alliés à la préservation des océans pourrait être une alternative.*

Introduction - The Galleon *Santíssimo Sacramento* (1668)

In the mid-1970s, the press reported the looting of archaeological remains, such as cannons and other artifacts, from a shipwreck by underwater fishermen (Book of the Ship *Gastão Moutinho* 2009). The Brazilian Navy, upon becoming aware of the situation, identified the wreck as the Galleon *Santíssimo Sacramento*, sunk in 1668, whose tragic history is well known in the city, Bay of All Saints. It was a flagship of the fleet of the *Companhia Geral do Comércio do Brasil*, which, according to documentary sources, carried around 600 people, including military personnel, sailors, and royal officials (Freire 1929).

Consequently, the Brazilian Navy, concerned with safeguarding valuable historical heritage from depredation, and in partnership with the Ministry of Culture and Education, hired archaeologist Ulysses Pernambucano de Mello Neto to coordinate a team of divers from the Navy to record the vessel and recover several artifacts that were vulnerable to looting during the summers of 1976 and 1978. The collection is currently curated in two Navy Museum collections, and its tragic maritime history attracts tourists to these exhibitions and to visit the wreck site through recreational diving.

Interpretation of the Data According to the Muckelroy Model

Since the creation of the model of wreck site formation processes by Muckelroy (1978), other analyses have followed, aimed at making maritime archaeology a multidisciplinary science to obtain more accurate understanding of underwater sites, whether in the context of the environment (Stewart 1999; O'Shea 2002; Bastida et al. 2010; Leino et al. 2011; Goulart 2014), or in the context of shipwrecks as an example of nautical culture (Gibbs 2006). However, Muckelroy (1978) developed a model about the deterioration process that is still valid, according to which observation and monitoring are also necessary to determine the most important threats to a specific location (Manders 2021).

Considering the state of underwater archeology in Brazil in the last 20 years, a period in which underwater archaeology was scientifically consolidated, a growing demand existed for the implementation of undertakings in the underwater environment. However, in Brazil, underwater cultural heritage (UCH) is the responsibility of the Ministry of the Navy, and Brazil does not follow the 2001 United Nations Educational, Scientific and Cultural Organization (UNESCO) Convention, whose recommendation prioritizes the preservation of underwater material heritage in situ. In addition, federal legislation prohibits the commercialization of (terrestrial) archaeological artifacts (Federal Law No. 3,924 of July 26, 1961) (Federal Law 1961), recognizing them as an asset of the union and as cultural heritage. But regarding material culture obtained from underwater heritage sites, Brazilian law allows their sale and public distribution (Federal Law No. 10,166/2000) (Federal Law 2000). Thus, one of the objectives of this research is to assess the degree of conservation of the wreck of the Galleon *Santíssimo Sacramento*, based on previous documentation (Mello Neto 1977, 1979; Guimartin, Jr. 1981; Torres 2016); letters and ordinances from the Governor General (Freire 1928a, 1928b, 1929); and given the current dispersion of structures and artifacts, under the Muckelroy (1978) model, to verify the possible natural and anthropogenic impacts on the site.

Geomorphological Aspects of the Wreck Area

Initially, as Muckelroy (1978) points out, the quality of preservation of the archaeological remains deposited in a wreck depends on coastal geomorphology, its hydrodynamic and environmental attributes, and marine ecology. In this instance, the coastal geomorphology of the wreck area in question is basically characterized

as the entrance to a large bay. The Bay of All Saints is characterized by a structure formed by a large accumulation of sand on the inner continental shelf of the city of Salvador, known as the Santo Antonio sandbank, an ebb delta built during the last 8 thousand years (8 ka). Its lateral position is the result of displacement by ebb currents. Additionally, this geomorphological scenario facilitates the occurrence of shipwrecks. The location of the sandbar has been reported by chroniclers and navigators since the 16th century as a danger to navigation (Mello 2016).

The area where the wreck is located is known from maps and engravings, which are available on the internet, as the scene of the war between the Iberian and Dutch powers in the first half of the 17th century. Examples include the engravings of the Albernaz family and unknown Dutch painters (Bandeira 2019). Besides the wreck of the Galleon *Santíssimo Sacramento*, other wrecks are also located at this entrance to the bay. They include *Cap Frio* (1908), *Germania* (1876), *Brittany* (1903), *Maraldi* (1875), *Reliance* (1884), and *Manaus* (1906) (Melo 2009; Torres 2016).

Results of the February 2020 Investigation

The delineation of the site is estimated at 45 meters (m) long and 20 m wide by means of the Pythagorean theorem, with a geometric correction of pending measurements, due to the depth of the wreckage at 32 m below the surface and the need for strict diving safety protocols. The wreck site was recorded from 03 February to 19 February 2020, through 63 dives totaling 32 hours and 03 minutes; 18 guns, 4 anchors, 1 mast cap, and the outline of the ballast line were recorded.

Comparison of the 1978 and 2020 Surveys

Figure 1 presents data from the 2020 investigation and Figure 2 shows the site map recorded by Mello Neto's team (1979). Comparing the old site map (Figure 2) and the new documentation (Figure 1), it can be observed that some areas remain as originally recorded, some areas have been modified, and some positions of the original record have been questioned due to the documentation activities that took place in 2020. First, the blue circles are the intact remains and correspond to the survey prepared by Mello Neto's team in the 1970s. The yellow circles represent the questionable positions of the pieces originally recorded, the pink circle are the observations from Torres´s research (2016), and the green circles

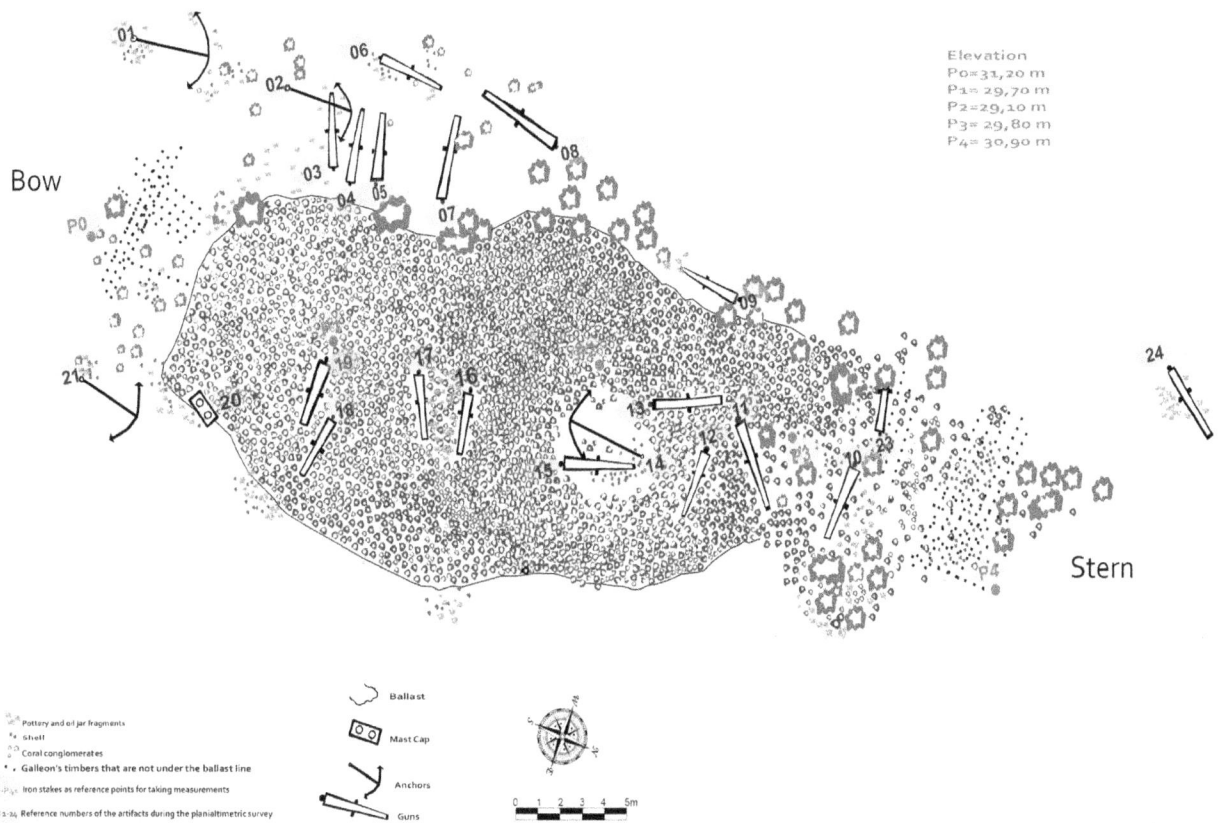

Elevation
P0=31,20 m
P1=29,70 m
P2=29,10 m
P3=29,80 m
P4=30,90 m

Bow

Stern

Pottery and oil jar fragments
Shell
Coral conglomerates
Galleon's timbers that are not under the ballast line
Iron stakes as reference points for taking measurements
Numbers 1-24 Reference numbers of the artifacts during the planialtimetric survey

Ballast
Mast Cap
Anchors
Guns

0 1 2 3 4 5m

FIGURE 1. Plan map showing new archaeological documentation of the Galleon *Santíssimo Sacramento* shipwreck (1668). (Figure by the author, 2023; magnetic north represented.)

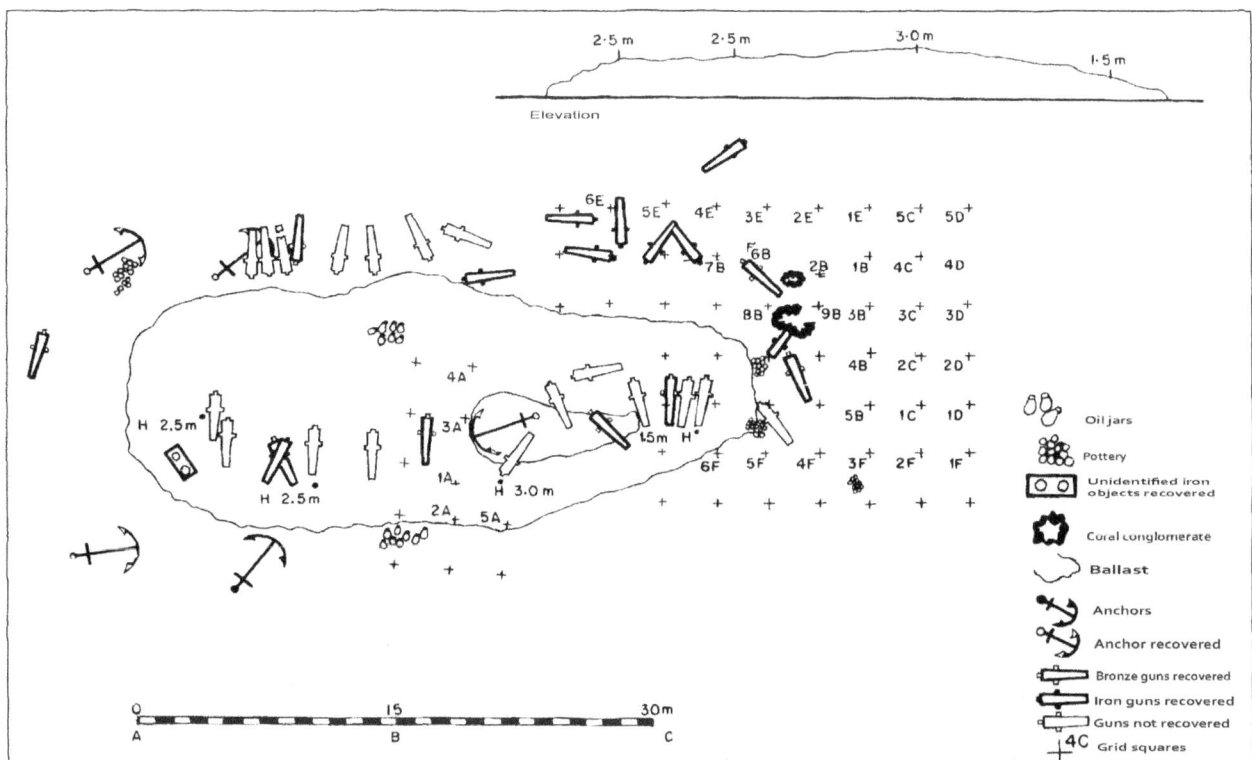

2·5 m   2·5 m   3·0 m   1·5 m

Elevation

Oil jars
Pottery
Unidentified iron objects recovered
Coral conglomerate
Ballast
Anchors
Anchor recovered
Bronze guns recovered
Iron guns recovered
Guns not recovered
Grid squares

0   15   30m
A   B   C

FIGURE 2. The shipwreck of the Galleon *Santíssimo Sacramento* shipwreck (1668) off Brazil, as recorded by Mello Neto in 1979. (Adapted from Mello Netto 1979.)

are the wooden remains more easily exposed, that are not under the ballast line (which line art represents the ballast mound). It is uncertain if the timbers mean the wooden ship structure is present or these are simply partial hull remains. The red circles indicate an area of the site, which is disturbed and altered between 2019 and 2020, the period in which the videos of the wreck were produced. These recent disturbances are possibly from damage caused by anchors during local tourist visits and/or fishing activities.

It is important to clarify that during the interpretation of the resulting data under the Muckelroy model, some information from the wreck site is the result of previous research. In addition to information from the research by Mello Netto's team (1977, 1979), the video recordings from the years 2015 and 2016, and some information from the *Santíssimo Sacramento* shipwreck that the author of this text will be citing were produced by Professor Rodrigo Torres (Centro Universitario Regional del Este, Uruguay) (East Regional University Center, Universidad de la República Uruguay), when he was preparing the material for his postdoctoral research, *Projeto Observabaía: Patrimônio Cultural Subaquático da Baía de Todos os Santos, Relatório Parcial* (2016) (Observabaía Project: Underwater Cultural Heritage of the Bay of All Saints, Partial Report), for the Observabaía Project (2023), a research program from the Federal University of Bahia (UFBA). Moreover, his information about the shipwreck of *Santíssimo Sacramento* can also be reviewed on the website The Nautical Archeology Digital Library (NADL), in collaboration with researcher Maria João Santos and the author of this text (NADL 2023).

Interpretation of the Data According to the Muckelroy Model

During the height of New Archaeology theoretical discussions, the premise arose that the archaeological record contains a sample of material traces from the past that is far from complete (Trigger 2004). In the United States, Schiffer proposed that archaeological data consisted of artifacts found in static relationships produced by cultural systems and subject to noncultural processes, composing a functionally integrated system (Schiffer 1976). In England, under the influence of David Clarke's knowledge of the importance of locational analysis around an archaeological site and general systems theory, Muckelroy's flowchart (Muckelroy 1978:158) illustrates the archaeological site of the shipwreck as an entrance to a closed system, and the exits from that system are the

transformations that the site undergoes through extraction filters and mixing devices after the shipwreck event. Whether from cultural or noncultural interventions, the filters extract the material from the assembled ship, and the mixing devices reorganize the contextual patterns of deposited archaeological remains.

In the case of this research, it is emphasized that the model is a particularistic perspective in which the goal is to expose the interpretations of the sinking process through observations in loco and the information produced prior to this research, as shown above, referring to Galleon *Santíssimo Sacramento*. Thus, the current results from the artifact analysis are a general understanding of the various degrees of knowledge that the different outputs reveal through the stages of transformation of this archaeological site, from the occurrence of the shipwreck until it was found by Mello Neto's team, after the excavations carried out in the 1970s and 1980s, and even in recent research between 2015 and 2020.

The Galleon

Santíssimo Sacramento (500 tons) was originally built between 1650 and 1651 for the fleet of the *Companhia Geral do Comércio do Brasil* in the city of Porto, northern Portugal, probably designed from the plans of Francisco Bento (Esparteiro 1976). According to letters from the Governor General Alexandre Souza Freire, the galleon brought with it the peace of Castella and reinforcement of ammunition for a possible attack by the enemy (Holland). The ship carried between 400 and 500 people (Freire 1929). *Santíssimo Sacramento* was armed with 52 or 53 guns at the time of the sinking; 26 were bronze cannons (Guilmartin, Jr. 1981). However, these were very old cannons. Some of them were Portuguese cannons, which were seized by the British and then resold to Portugal (Brown 2005).

Process of Wrecking

At dusk on 05 May 1668, when trying to enter the Bay of All Saints along with nine other ships of the Armada amidst strong southern winds, *Santíssimo Sacramento* failed to reach land. The ship collided with the crest of the Banco de Santo Antonio, which then fired its cannons to signal imminent danger. However, due to the storm, no vessel was able to leave the port to help the sinking ship (Freire 1929).

According to letters written by the Governor General Freire to the Governor of Rio de Janeiro, D. Pedro Mascarenhas, and to the Ombudsman of Rio de Janeiro, Diogo Carneiro de Fontoura, between 400 and 500 people on board drowned, including Navy General Francisco Correa da Silva; only 70 individuals survived (Freire 1929). Comparing the site plan maps of 1979 and 2020, the position of the anchor and the gun on the deck, it appears there was no time to use it to anchor. According to the captain of the Navy diving team during Mello Neto's research, it was observed by the divers that some stern guns were buried under the seabed, leading to the assumption that an opening occurred in that area during the wreck (Guimartin, Jr. 1981). This information can also be corroborated by the position of an iron cannon further from the wreck, on the extreme side of the stern. However, in Mello Neto's record (1979), this cannon is positioned on the opposite side (extreme side of the bow), due to a recording error.

Material which Floated Away

According to the tragic maritime news from the time of the shipwreck, as reported above, the event took place in the midst of a storm. Therefore, it is likely that due to the rough seas, some parts of the ship's hull may have broken off and floated away. It is known through historical records, an ordinance record that is quite fragmented by moths, in which the Governor-General gives orders to the Sergeant Major to go to the Morro to see the *Santíssimo Sacramento's* hull on 09 November 1668, which can be interpreted to regard the wreckage of the hull stranded near a hill (Freire 1928).

Unfortunately, the document does not specifically mention what exactly the galleon was deposited with in this location, or the way in which the site settled on the seabed. Noting the position of artifacts recorded in Figures 1 and 2, it is imagined that it was dragged down by a combination of ballast weight, cargo, and water intake. In this process, what is deposited at the foot of the hill could be parts of the hull or dead wood from the mast. The mast broke from the violence of the storm, and due to that the wood was not likely waterlogged enough to sink. The marine current during the storm deposited this wood where it was found.

Salvage Operation

Regarding the material that would be actively salvaged at the time, the evaluation of this extraction device from the shipwreck site formation process depends on some aspects, such as the site of the wreck, whether there is evidence of historical records about salvage activity and of the material found at the wreck site (Muckelroy 1998:275).

The wreck site is located at a depth of 33 m below water, and almost two miles from the coast. The location limits the application of a rescue operation, such as dragline, and the carrying out of rescue free dives for survivors. Next, no evidence from the historical record indicates whether a rescue operation was attempted. In this sense, it is assumed that the artifacts collected during the excavation activities of Mello Neto's team between the summers of 1976 and 1978 corroborate the news from the General Governor, as shown in the topic *Process of Wrecking* (above), that it was not possible to rescue people from the wreck at the time of the sinking. In the artifact collection is a range of material types, including naval artillery, nautical instruments, crew cargo, and nautical work, including ferrous and non-ferrous metal, glass, materials of organic origin, lead seals, earthenware, coins, and ceramics (Mello Neto 1977, 1979).

Disintegration of Perishables

According to the remains highlighted in blue circles in Figure 3, these artifacts remain in the original position of Mello Neto's record, their positions highlight the process of disintegration of perishables as characterized by the disintegration of the ship's timbers that were above the ballast line. These probably disintegrated by the natural extraction devices, as the Brazilian tropical salt waters provide low salinity and oxygen, an easy environment for the deterioration of submerged wood. The disintegration process can be observed by the position of some guns, anchors, and mast cap.

Initially, the cannons muzzles are facing the outside of the ship (see respectively cannon numbers 3–5, 7, 10, and 16–19, and 23 on Figure 3) both on the starboard and port sides. Together, with these cannons, the anchors on the seabed and the anchors with the cannon in the center of the site are shown. Guilmartin, Jr. (1981), when analyzing the characteristics of the guns recovered by Mello Neto's team and the 1979 record, observed the distribution of wreckage, and suggested that the ship stopped on its starboard hull on a relatively balanced keel. This arrangement is clear in the arrangement of anchors and guns: the guns originally lay in two irregular parallel lines. This position of the guns, in turn, indicates their locations on a horizontal plane of the ship before it descends. The gun lines curve inwards at the extreme stern, just enough to suggest that the two guns on the nearest opposing lines were in a strategic location on the stern, mounted side-by-side to fire backwards from either side of the rudder. The gun lines are less regular aft, where the hull and superstructure would have been

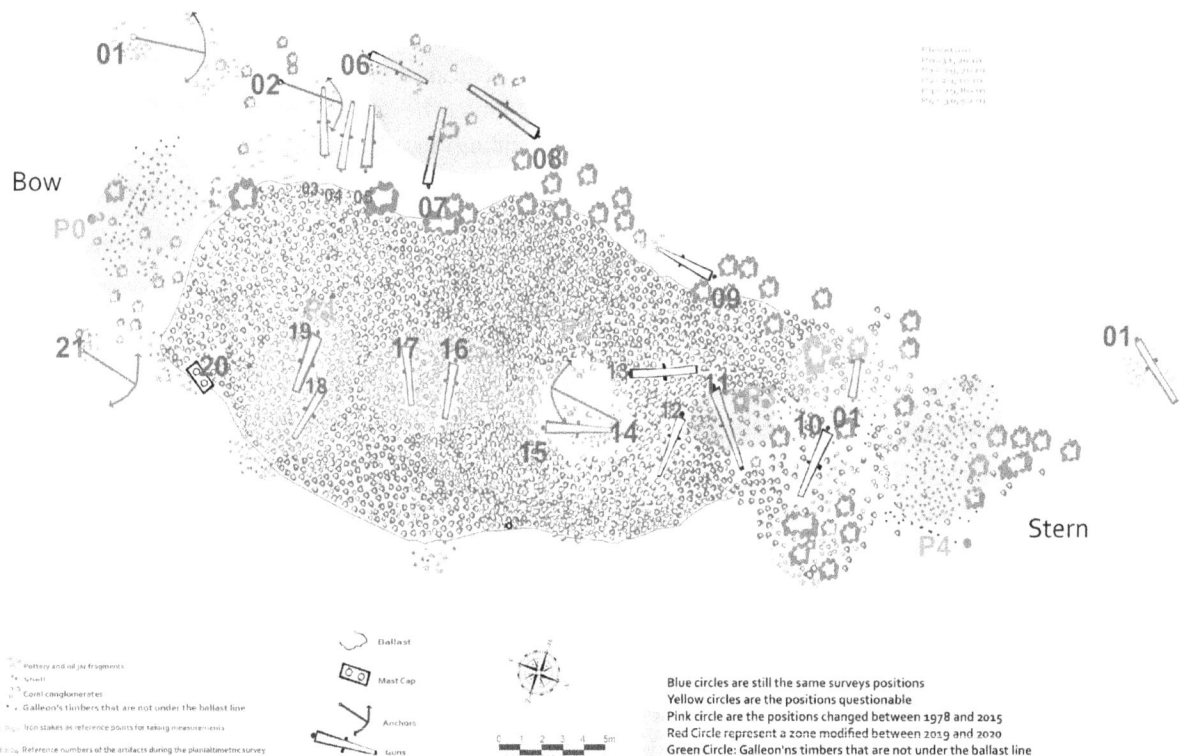

FIGURE 3. Colored circles over artifacts recorded during the 2020 survey that represent areas of stability, possible recording errors, and changes between the Mello Neto (1979) record and the current investigation. (Figure by the author, 2023).

deeper, leaving a greater mass of rotten wood to clutter the gun ranks on their slow journey to the bottom as the wreckage decayed (Guimartin, Jr. 1981).

At the time of the 1979 record, some photos of the activities of the team of divers were recorded, and there is an image that records the position of one of the anchors with its shank over one of the cannons, located in the bow, starboard side. It is understood, therefore, that the anchor fell after the cannons, due to the disintegration of the wooden hull. It is a position that still holds true. On the port side of the bow, next to an anchor and close to one of its nails, rests a mast cap from one of the masts. This must have broken with the impact of the ship, and subsequently disintegrated. The piece is made of iron and is 1.5 m long. In addition, the cannon and the anchor close to the center of the site (see numbers 14 and 15 on Figure 3) suggest the central location of the deck, which landed on top of the ballast pile. The position of most of the ballast indicates that it was in the hold of the ship at the time of the sinking, although there is a disturbance on the starboard and stern sides, whose observation is described in the heading *Characteristics of the Excavation* (below).

Movement of the Seabed

This topic explores the conditions of the ocean that have influenced the shipwreck. Although site depth protects the site from major hydrodynamic impacts, unlike shipwrecks located at shallower depths, the shipwreck is located near the entrance to a bay, which in turn is an area for tidal exchange, as well as of the actions of waves and currents. Furthermore, it is important to consider that the wreck is 1.5 kilometers (km) away, towards the coast, from one of the main outfalls of the city, the Rio Vermelho outfall; few published studies exist about the outfall regarding whether the waste can influence the quality of the water (Melo 2009).

In turn, the movement of the seafloor impacts the preservation of the site and is responsible for transporting sediments and depositing objects. This action will have a direct impact on the conservation of the entire site due to the shape resulting from the wrecking process on some pieces, as observed in the section *Disintegration of Perishables* (above). In this case, the amount of ballast reaching 3 m in height, was deposited mostly in the central area of the site. In this sense, the currents come and go according to the tidal regime (sometimes ebb, sometimes flood), in which the ebb tide stands out more (previously discussed above under the heading *Some Geomorphological Aspects of the Wreck Area*), the sediments transported tend to accumulate aft and along the starboard side of the site.

During visits to the site between 2019 and 2020, an accumulation of sediments was observed at cannon number 8 (Figure 3), which is located ahead of midship, on the starboard side of the site. This resulting accumulation is also due to the height of the ballast, which hinders the speed of transport energy, causing the archaeological remains that are found at the top to receive less sediment than those that are deposited on the seabed.

Furthermore, a few bottles of beer and lavender perfume are scattered over the site. According to a personal conversation with a master of the *Rio Vermelho Z-1* fishermen's colony, these perfume bottles come from a deposit located 5 km from the emissary where offerings to the African orixá Iemanjá (annual festival to Brazil's Goddess of the Sea, Iemanjá) are usually launched every year on February 2, (Master Luiz 2022 pers. comm.). This is an important day for the city, which also receives a large contingent of tourists to participate in the festivities. Therefore, due to the action of surface waves and currents, when the fishermen throw the offerings into the sea, part of them naturally float and sink along the route, guided by the direction of the current. Since the 1979 investigation of the site (Mello Neto 1979), objects deposited because of human intervention on the site may also be present. This is the case of a piece of rope located at the stern, which was covered by a thin layer of surface sediment when suspended to record the most exposed woods. Next, two observations still do not have concrete answers. These are regarding the growth of corals on ferrous artifacts, and a recurring exposure of the wooden hull that is not located under the upstream deposition of the ballast. First, a rapid growth of coral on ferrous objects was noted between 2019 and 2020; between 2015 and 2016, the coral growth was present, but in a discreet way and more concentrated on the pieces deposited on the seabed, causing this growth to spread across all artifacts deposited at the site in recent years. Then, the 2019 and 2020 videos show a greater exposure of the wooden hull that is not under the ballast mound, something that was discreet or almost imperceptible in the 2015 and 2016 videos (see the green circles on Figure 3).

When dealing with the issue of transport and suspension of sediments, it is known that seasonality affects the transport regime. In winter periods, the suspension is usually denser and slower, and in the summer, visibility lighter. However, one should also pay attention to the influence of air masses that come from the outside, which depending on the year and season, come with more intensity, influencing the energy of sediment transport on the seabed. In 2019, winter weather recorded large and overwhelming waves on the coast. This may have contributed to a greater exposure of the wooden hull in the area of the bow and stern that are not under the ballast mound. The archaeological remains that are not at the top of the ballast mound received a greater amount of sedimentation.

The videos recorded a period of five years of site visits, but they were not standardized according to seasonal periods. For this reason, the last two observations described above still require more analysis. The visits to the *Santíssimo Sacramento* site in 2015 and 2016 were carried out respectively in spring (September) and summer (March), and those in 2019 and 2020 were carried out respectively in the months of May, June, August, September, November, December, January, and February, in the seasons of autumn, winter, spring and summer.

Characteristics of the Excavation

Mello Neto (1979) described the excavation methodology applied at the wreck site: "Planimetric and photographic surveys were carried out, and a 3 × 3 m grid was made before extracting the artifacts, initially with manual collection of material and later with suction equipment, namely an aqualift" (Mello Neto 1979:211). In this sense, the plan mapping was carried out to accurately position the "notable" pieces that made up the ship, such as cannons, anchors, and the mainstay, which Mello Neto records as undefined; and the ballast and ceramic sets, which were at risk of being removed by treasure hunters. As for the smaller pieces, which make up the rest of the collection, it is not known where exactly the artifacts were collected in these 3 × 3 m grids. There is also no information about the method of collection, and the depositional context within the wreck site.

Furthermore, according to a personal report by the captain of the vessel that recovered the archaeological remains, Rocha de Oliveira, as reported to Professor Rodrigo Torres, "during two days of diving, it was first thought they would remove all the ballast and reassemble it on land, maintaining relative positions. However, the idea was aborted when realizing that it would take too long" (Rocha de Oliveira 2015, pers. comm.). Based on this information, the outline of ballast is currently disturbed in some areas and does not follow an ordered pattern as shown in the 1978 mapping. Another result of this excavation, but which could also be the result of later investigations, are the tiny fragments of ceramics and crockery currently scattered throughout the site. After research by Mello Neto's team, salvage licenses were issued by the authorities of the Brazilian Navy to private exploration companies, which caused great disturbance to the site without the proper techniques for registering or publishing the findings.

Observed Seabed Distribution

Comparing the similarities and differences between the site maps of the *Santíssimo Sacramento* from 1979 and 2020, analysis was made of precisely what changed after Mello Neto's investigation. The differences lie mainly in the starboard side of the ship and the stern area, as well as material in different sizes dispersed throughout the site. Cannons, ballast stones, pieces of wood, undefined concretions and thousands of fragments of pottery and crockery in tiny sizes are scattered across the site. In addition to the differences in the positions of the artifacts among the plan maps, described in the steps of the process of this wrecking described above, some visible changes were also observed on the videos produced during the five years of research (2015–2020).

This comparison began with observing the position of the cannons that lie in the submerged bed on the starboard side (see cannon numbers 6–8 on Figure 3). Between Mello Neto's record and the observations made by Torres in his postdoctoral project (2016), one can see a change in the cannons in this area. These positions are most likely related to human intervention to the site after Mello Neto's first documentation of the site, as described in the *Characteristics of the Excavation* section above. Then, another point that deserves to be highlighted here is the issue of ceramic and pottery fragments, as well as the outline of the ballast line. Regarding the latter, if one circles the site, it is soon evident that on the starboard side, the ballast line is not deposited in the same way as on the port side; it is more dispersed. In the same way, the ceramic and pottery pieces, quite fragmented throughout the site, are deposited around the cannon that is on the extreme side of the stern. When observing the outline of the ballast line and the ceramic fragments, such mixing devices appear to be the result of the intervention of the Navy divers' team, which were intensified with subsequent investigations.

Regarding the observations made from the videos, notable changes were observed in the stern area, around gun 12, between the 2019 and 2020 videos. This gun was characterized by a coral growth close to its breech; over it, there was a round concretion. In the last videos of 2020, it is visible that the concretion was removed, and the coral knocked down, leading to the belief that it may have been the result of anchoring on the wreck site. And finally, another notable feature is the growth of corals on ferrous objects over the five years of visits to the site, which was described in the *Seabed Movement* section above.

Future Perspectives

Even though Brazil does not follow the 2001 UNESCO Convention, and the submerged heritage in Brazil still suffers from ongoing depredations, the author believes that there are alternatives to strengthen protection of UCH in Brazil. One of them is through teaching and maintaining an ongoing dialogue between archaeologists, divers, and diving schools, a practice that has given fruitful results over 20 years (Duran et al. 2014). Publication of the results of investigations like the analysis discussed herein is part of that dialogue. The other alternative is to implement a policy for the preservation of underwater sites to strengthen sustainable tourism, measures which are already being put into practice, in the interest of maintaining oceanic biodiversity of the Bay of all Saints (Hatje and de Andrade 2009). Examples are the *Parque Marinho da Barra* (Marine Park of Barra) and the *Parque Marinho Cidade Baixa* (Marine Park of Lower City), both located on popular swimming beaches, whose areas delineate important underwater archaeological sites that are often visited by divers due to their good diving conditions, such as the shipwrecks *Maraldi* (1875), *Germania* (1876), *Brittany* (1903), and *Blackadder* (1905). These are initiatives of social interest that had the support of the Municipal Tourism Secretariat, from the Federal Institute of Bahia, and even the Regional Superintendence of the National Historical and Artistic Heritage Institute of Bahia, the latter collaborating in the creation of the Marine Park of Barra. Such initiatives strengthen the preservation of Brazil's underwater cultural heritage.

Acknowledgements

I would like to thank those who helped make this research project a success: Dr. Gilson Rambelli (Federal University of Sergipe-UFS); Dr. Leandro Duran (UFS), Professor Rodrigo Torres (Centro Universitario Regional del Este, Uruguay); Dr. Flávio Calippo (Federal

University of Piauí - UFPI); my professional colleagues Marcus Davis Braga (UFCE), Júlio César Marins (UERJ), the NAUI diving instructors and rebreathers Alvanir Oliveira (Jornada) and Oswaldo Del Cima; and finally, the Galeão *Sacramento* diving school instructors Bruno Rocha, Juvenal Barreto, and assistant instructor Luni Otashima. Thanks also to the professors who have supervised the current stage of research, Professor Filipe Castro from the Center for Functional Ecology - Science for People and the Planet (CFE), R&D unit of the Department of Life Sciences at the University of Coimbra and at the Institute of Contemporary History of the NOVA University of Lisbon, and Professor José Maria Landim Dominguez, geologist at the Federal University of Bahia in Coastal and Sedimentary Geology.

References

BANDEIRA, BEATRIZ BRITO DE FERREIRA
2019   O Galeão *Sacramento* e a Rota do Açúcar: Por uma Arqueologia da PaisagemThe Galleon *Sacramento* and the Sugar Route: An Archaeology of the Landscape). In *Arqueologia de Engenhos*, 2019, Recife. Arqueologia de Engenhos. Organizadores: Oliveira,C., Ghethi, N. C., Allen, S. J. Vol.1 Paisagens e Pessoas. Recife: Editora UFPE. v. 01.

BASTIDA, R., D. ELKIN, AND M. GROSSO
2010   Enfoques Interdisciplinarios para el Estudio de Procesos Naturales de Formación de Sitios Arqueológicos Subacuáticos: Investigaciones en el Marco del Proyecto Swift (Provincia de Santa Cruz, Argentina), (Interdisciplinary Approaches for the Study of Natural Processes of Formation of Underwater Archaeological Sites: Investigations in the Framework of Proyecto Swift (Province of Santa Cruz, Argentina). In *Arqueología argentina en los inicios de un nuevo siglo*. Editors Oliva, F.; Grandis, N.; Rodríguez, J. (de). Publicación del XIV Congreso Nacional de Arqueología Argentina (Rosario, Septiembre 2001). Rosario: Laborde Editor, Tomo III, p. 269–283.

BOOK OF THE SHIP GASTÃO MOUTINHO
2009   Navy Archive, in the Special Documents Division, located in the Directorate of Navy Historical Heritage (DPHDM), in the Command of the 1st Naval District, in Rio de Janeiro City, Brazil.

BROWN, R.
2005   Seis Canhões do Século XVI Provenientes do *Santíssimo Sacramento*: Uma Reestimativa (Six 16th-Century Cannons from the *Santíssimo Sacramento*: A Reassessment). *Navigator, Rio de Janeiro* 1(2):21–34.

DURAN, LEANDRO D., PAULO F. BAVA-DE-CAMAGRO, FLÁVIO R. CALIPPO, AND GILSON RAMBELLI
2014   *Educando Embaixo D'Água: Uma Perspectiva Histórica do Ensino de Arqueologia Subaquática no Brasil (1992–2014) (Educating Underwater: A Historical Perspective on Teach Underwater Archaeology in Brazil (1992–2014)*. Revista Habitus, PUC-GO.

ESPARTEIRO, ANTÔNIO M.
1976   *Catálogo dos Navios Brigantinos (1640–1910) (Catalog of Brigantine Ships)*. Centros de Estudos da Marinha, Lisboa; Três Séculos no Mar, v. 2. 2 parte, Ministério da Marinha, Lisboa,

FEDERAL LAW
1961   Federal Law No. 3,924, July 26, 196, <http://portal.iphan.gov.br/uploads/legislacao/Lei_3924_de_26_de_julho_de_1961.pdf>. Accessed 04 April 2023.

2000   Federal Law No. 10,166, December 27, 2000, <https://www2.camara.leg.br/legin/fed/lei/2000/lei-10166-27-dezembro-2000-356794-publicacaooriginal-1-pl.html>. Accessed 19 May 2023.

FREIRE, ALEXANDRE DE SOUSA
1928a   *Carta que se escreveu ao governador do Rio de Janeiro, Dom Pedro Mascarenhas (Letter written to the governor of Rio de Janeiro, Dom Pedro Mascarenhas)*. In DOCUMENTOS Históricos da Biblioteca Nacional. Rio de Janeiro, 1928. v. 6, p. 92. <https://www.gov.br/fundaj/pt-br/composicao/dimeca-1/biblioteca/acervos/inventarios-documentais-e-indices/documentacaohistoricabn.pdf>. Accessed 08 April 2023.

1928b   *Carta que se escreveu ao provedor do Rio de Janeiro Diogo Carneiro da Fontoura (Letter written to the provider of Rio de Janeiro Diogo Carneiro da Fontoura)*. In DOCUMENTOS Históricos da Biblioteca Nacional. Rio de Janeiro, 1928. v. 6, p. 97. <https://www.gov.br/fundaj/pt-br/composicao/dimeca-1/biblioteca/acervos/inventarios-documentais-e-indices/documentacaohistoricabn.pdf>. Accessed 08 April 2023.

1929   *Carta que se escreveu ao governador da capitania de Pernambuco, Bernardo do Miranda Henriques (Letter written to the governor of the captaincy of Pernambuco, Bernardo do Miranda Henriques)*. In: Documentos Históricos da Biblioteca Nacional Rio de Janeiro v. °, p. 204. <https://www.gov.br/fundaj/pt-br/composicao/dimeca-1/biblioteca/acervos/inventarios-documentais-e-indices/documentacaohistoricabn.pdf >. Acessed 08 April 2023.

GIBBS, MARTIN
2006   Cultural Site Formation Processes in Maritime
        Archaeology: Disaster Response, Salvage and
        Muckelroy 30 Years on. *The International Journal of
        Nautical Archaeology* 35(1):4–19.

GOULART, LUANA B. G. J.
2014   Processos de formação arqueológicos de sítios de
        naufrágios: uma proposta sistematica de estudos
        (Archaeological formation processes of shipwreck
        sites: a systematic proposal of studies). (Dissertação
        de Mestrado em Arqueologia), Programa de Pós-
        Graduação em Arqueologia da Universidade Federal
        de Sergipe, Laranjeiras. <https://www.researchgate.
        net/publication/282018862_Processos_de_
        formacao_arqueologicos_de_sitios_de_naufragios_
        uma_proposta_sistematica_de_estudos>. Accessed
        20 May 2023.

GUIMARTIM, JR., JOHN, F.
1981   Os canhões do *Santíssimo Sacramento* (The
        Cannons of the *Santíssimo Sacramento*). Navigator,
        subsídios para a história marítima do Brasil.
        Serviço de Documentação Geral da Marinha,
        No 17, jan-dez de 1981. pp. 3–44. https://www.
        portaldeperiodicos.marinha.mil.br/index.php/
        navigator/article/view/1070. Accessed 20 May
        2023.

HATJE, VANESSA, AND JAILSON B. DE ANDRADE (EDITORS)
2009   *Baía de todos os santos: aspectos oceanográficos
        (Santos Bay: Oceanographic Aspects).* Editora da
        Universidade Federal da Bahia, Salvador, Brazil.

LEINO, M., A RUUSKANEN, J. FLINKMAN, J. KAASINEN, U.
KLEMELÄ, R. HIETALA, AND N. NAPPU
2011   The Natural Environment of the Shipwreck
        Vrouw Maria (1771) in the Northern Baltic Sea:
        An Assessment of her State of Preservation. *The
        International Journal of Nautical Archaeology*
        40(1):133–150.

MANDERS, M.
2021   Unit 4: The Underwater Archaeological Resources.
        In *The UNESCO Training Manual for the Protection
        of the Underwater Cultural Heritage in Latin
        America and the Caribbean*, pp. 79–96. United
        Nations Educational, Scientific and Cultural
        Organization: Amersfoort, The Netherlands.
        https://unesdoc.unesco.org/ark:/48223/
        pf0000375747. Accessed May 2023.

MELLO, ANA CLARA C.
2016   O Banco de Santo Antônio: um estudo de sísmica
        de alta resolução em um delta de maré vazante
        localizado na entrada de uma grande baía tropical,
        costa leste do Brasil (The Santo Antônio Bank: A
        High-Resolution Seismic Study in an Ebb Tide
        Delta Located at the Entrance of a Large Tropical
        Bay, East Coast of Brazil. Federal University of
        Bahia, Salvador, Brazil. Dissertation, <https://
        repositorio.ufba.br/handle/ri/21571?locale=en>.
        Accessed 20 May 2023.

MELO, LIZANDRA C. F.
2009   Usos múltiplos e proposta de disciplinamento da
        plataforma continental em frente ao município
        de Salvador – Bahia (Multiple Uses and Proposed
        Regulation of the Continental Shelf in Front of the
        Municpality of Salvador – Bahia). Dissertação de
        mestrado. Universidade Federal da Bahia Salvador,
        Brazil.. Dissertation, https://repositorio.ufba.br/
        handle/ri/16222. Accessed 20 May 2023.

MELLO NETO, ULISSES P.
1977   O Galeão *Sacramento* (1668): Um naufrágio
        no século XVII e os resultados de uma pesquisa
        de arqueologia submarina na Bahia/Brasil (The
        Galeão *Sacramento* (1668): A Shipwreck of the
        17th Century and the Results of an Underwater
        Archaeological Survey in Bahia, Brazil) . "Revista
        Navigator", n. 13. Marinha do Brasil: Rio de
        Janeiro.

1979   The Shipwreck of the Galleon *Sacramento* 1668
        off Brazil. T*he International Journal of Nautical
        Archaeology and Underwater Exploration* 8(3):211–
        223.

MUCKELROY, KEITH.
1978   *Maritime Archaeology.* Cambridge: Cambridge
        University Press.

1998   The Archaeology of Shipwrecks. In *Maritime
        Archaeology: A Reader of Substantive and Theoretical
        Contributions.* edited by Lawrence E. Bablts and
        Hans Van Tilburg. Springer Science and Business
        Media: New York, pp. 267–290.

THE NAUTICAL ARCHEOLOGY DIGITAL LIBRARY (NADL)
2023   *Santíssimo Sacramento*, 1668. <https://shiplib.
        org/index.php/shipwrecks/iberian-shipwrecks/
        portuguese-merchantmen/santissimo-
        sacramento-1668/>. Accessed 08 April 2023.

OBSERVABAÍA PROJECT
2023   Observabaía Project. <https://observabaia.ufba.
        br/o-observabaia/>. Accessed 08 April 2023.

O'Shea, John M.
2002   The Archaeology of Scattered Wreck Sites:
        Formation Processes and Shallow-Water
        Archaeology in Eastern Lake Huron. *The
        International Journal of Nautical Archaeology*
        32(2):211–227.

Schiffer, M. B.
1976   *Behavioral Archaeology.* Academic Press: New York,
        New York.

Stewart, D.
1999   Formation Processes Affecting Submerged
        Archaeological Sites: An Overview. *Geoarchaeology:
        An International Journal* 14(6):565–587.

Torres, Rodrigo de O.
2016   Projeto Observabaía: Patrimônio Cultural
        Subaquático da Baía de Todos os Santos
        (Observabaía Project: Underwater Cultural
        Heritage of Todos os Santos Bay). Relatório Parcial
        2015. *Revista Navigator* 12(23):140–153.

Trigger, Bruce
2004   *A History of Archaeological Thought.* Cambridge
        University Press: Cambridge, England.

· · · · · · · · · · · · · · · ·

Beatriz Bandeira

Graduate Program in Archeology at the

Federal University of Sergipe (Proarq-UFS)

Salvador

Bahia, Brazil 40130-030

Maritime Survey Results of LaSoye Bay, Dominica

Marie Marenda, Megan Bebee

After the discovery of a potential 17th-century town site was uncovered on the beach of LaSoye Bay, Dominica, in the wake of Hurricane Maria in 2017, a project was designed to document the maritime culture of the area. The bay is one of the few harbors off the Atlantic coast offering relatively calm conditions for safe anchorage. The study incorporates a range of methods to characterize both the environmental and anthropogenic features of the bay and to understand how it has changed over time. Two field seasons have focused on the material remains, utilizing diver-led survey and a Garmin FishFinder, and this study examines the environmental history through sediment coring. The bathymetric survey results from 2022 were incorporated into a 3D-bathymetric model, which informs how ships navigated the bay. The study gives insight into the maritime trade on the isolated windward coast, and illustrates how use of the bay has transformed over time.

Após a descoberta de uma potencial cidade do século XVII na praia da Baía de LaSoye, na Domínica, na sequência do furacão Maria em 2017, foi concebido um projeto para documentar a cultura marítima da zona. A baía é um dos poucos portos da costa atlântica que oferece condições relativamente calmas para um ancoradouro seguro. O estudo incorpora uma série de métodos para caraterizar as característica ambientais e antropogénicas da baía e para compreender a sua evolução ao longo do tempo. Duas épocas de trabalho de campo centraram-se nos vestígios materiais, utilizando um levantamento conduzido por mergulhadores e um Garmin FishFinder, e este estudo examina a história ambiental através da recolha de sedimentos. Os resultados do levantamento batimétrico de 2022 foram incorporados num modelo batimétrico 3D, que informa sobre a forma como os navios navegavam na baía. O estudo fornece informações sobre o comércio marítimo na isolada costa de barlavento e ilustra como a utilização da baía se transformou ao longo do tempo.

Après la découverte d'un site potentiel d'une ville du 17e siècle sur la plage de la baie de LaSoye, en Dominique, à la suite de l'ouragan Maria en 2017, un projet a été conçu pour documenter la culture maritime de la région. La baie est l'un des rares ports sur la côte atlantique offrant des conditions relativement calmes pour un mouillage sûr. L'étude intègre une gamme de méthodes pour caractériser à la fois les éléments environnementaux et anthropiques de la baie et pour comprendre comment elle a changé au fil du temps. Deux saisons sur le terrain se sont concentrées sur les restes matériaux, en utilisant un des recherches avec plongeurs et un FishFinder de Garmin, et cette étude examine l'histoire environnementale à travers le carottage des sédiments. Les résultats du levé bathymétrique de 2022 ont été incorporés dans un modèle bathymétrique 3D, qui informe la façon dont les navires ont navigué dans la baie. L'étude donne un aperçu du commerce maritime sur la côte au vent isolée et illustre comment l'utilisation de la baie s'est transformée au fil du temps.

Introduction

The project explored the long-term human use of LaSoye Bay, Dominica. The remains of the early 17th-century port town were discovered during the aftermath of Hurricane Maria in 2017 eroding out of the beach. The bay, and the archaeological sites along its coast, offer unique insights into the changing landscape from prehistoric occupation into the modern era. Previous work in the area indicates this was a site of Indigenous occupation (Boomert 2009, 2011), in addition to the colonial port town (Hauser et al. 2019) and post-colonial structures present along the coast. This dissertation project was created to determine the extent of submerged and coastal remains of the anchorage, and aims to characterize the maritime landscape of the area from early human occupation to present. Archaeological and geological methods, including diver-led survey, targeted surveys using a Garmin UHD Echosounder, bathymetric mapping and modeling, recording of shoreline erosion, and sediment sampling with cores have been employed to capture the anthropogenic and environmental historical shifts of the anchorage. This paper covers the first two field seasons, which primarily focused on documenting archaeological remains related to the maritime context in, and around, the bay and establishing an environmental context for the marine environment.

The bay is located on the Atlantic-facing side of Dominica and looks directly out at the island of Marie Galante. On clear days, one can see Les Saintes, the outlying islands of Guadeloupe to the north. Situated along the Dominica Passage, an active trading channel from the 16th to 19th centuries, this is one of the few harbors on the windward coast offering relatively calm conditions and protection from the winds for safe anchorage. This small anchorage is marked on a 1771 map recovered from local archives (Bowen 1790) indicating its prominence during the height of European conquest in the Caribbean. The coastline is comprised of three separate beaches, which are marked on the map in Figure 1. Woodford Hill Beach (Section A) stretches from *Point La Soie* to the rocky outcrop. Woodford Hill Beach (Section B) is the primary public beach and the location of the 17th-century port town. Little Turtle Beach (Section C) is the beach furthest north along the bay and presents an entirely different environment than the Woodford Hill beaches. This area is subject to heavy wave action and is open to the Atlantic, without the protection of the bay that is granted to the other two beaches.

FIGURE 1. Map of LaSoye Bay, facing true north. (Figure created by the author, 2022.)

Extant remains around the shore include a bollard to tie off boats, the remnants of a seawall, ruins of a lime factory, and an abandoned fishing structure from the 1950s. Furthermore, terrestrial archaeology from around the area indicates various periods of human occupation from pre-contact forward. Although this is a relatively small (approximately 1,547.21 square meters [m^2]) bay, it has actively been used over time. Up to the 1950s, the bay was sheltered from the southeast by a headland, *Pointe La Soie*, providing protection from winds and from sight by anyone entering the southern Caribbean from the Atlantic. Ships coming from Europe or Africa using the "southern passage" would enter the Caribbean through the passage between Marie Galante, Dominica, and Guadeloupe. Traditionally, ships would sail around the northern point of Dominica and make harbor on the calm deep-water port of Prince Rupert Bay, Portsmouth, where fresh water and provisions would be restocked after the long Atlantic crossing. However, Dominica's unique position as a neutral territory into the 17th century, controlled by the Indigenous Kalinago, created a pocket for settlers from other islands coming over to Dominica to settle. By the 1730s, a town was developing in Roseau for ease of trade with neighboring Martinique. Additionally, settlers spread out along the coasts at bays for anchoring "such as Calihaut, two spots in the north including LaSoye and Grande bay in the South" (Borome 1967:19).

The relative lawlessness during the early to mid-1700s made Dominica ripe for illegal activities and trade. In 1766, the British Parliament passed the Freeport Act opening up two ports in Dominica to trade with the French to increase the British-produced market in Guadeloupe and Martinique, and have the French ship agricultural produce to Britain through these ports (Hauser 2011). However, an analysis of 18th-century ceramics from two plantations on Dominica indicates that free and enslaved residents sought goods from inter-island trade networks (Hauser 2011), which were likely established prior to the imperial authorization and then continued into the 19th century under Britain's attempt to increase their own market. The mountainous topography of Dominica creates small, isolated pockets, particularly along the windward coast where no large government-controlled ports exist, meaning that activities could take place that were far less regulated. This also meant even fewer records of activity in bays such as LaSoye exist for much of this period.

Activity has never ceased in this bay, further adding to the intriguing history of the area. Local historian,

Lennox Honeychurch, has a grandfather, Lennox Napier, who recorded a controversy between the landowner of Woodford Hill Estate and the fishermen who wanted to use the bay and build a fishery in 1937. Stanley Bruno says:

> It was a shipping port. I had a contract to load and unload the government steamers the Yare and 'Montedoro'and Lefear and the Mermaid which called regularly. I then at the time had two boats and 6 men working with me [yare big steamboat] they would go back and forth with the small boats to load cargos]. Schooners and sloops called regularly in the old days and there were lots of boats regularly kept on the foreshore. There was a wharf there, which was used by all the neighboring estates for shipping sugar and other produce and receiving supplies. There was also a market on the foreshore. James O'Brien of Wesley was one of the estate owners who used Woodford hill bay for a shipping port. Since I can remember and up to the present date, boats have been using Woodford hill bay as a port. A big boat came to unload colas [the car for building the road]. (Lennox Honeychurch pers. comm. 2022)

The road mentioned above was paved in 1915. These accounts refer to conditions during World War I and prior. This offers valuable insight into activities in the bay during the first few decades of the 20th century.

Preliminary Field Season 2019

The first fieldwork was conducted in 2019 as an exploratory season to test the viability of conducting dissertation research on the underwater remains within the bay. The survey strategy involved a combination of snorkel and dive transects. It was a quick season, lasting only two weeks. Initial surveys began on snorkel transects close to the terrestrial site and public beach. Visibility was limited due to high turbidity, but close to the terrestrial site was evidence of coastal erosion, with pipe stems and pottery fragments being found in the shallow waters offshore. Working into water deeper than three meters (m) proved fruitless given the low visibility.

The next phase of the snorkel survey took place to the north in Little Turtle Bay. After talking to local historian, Lennox Honeychurch, who described a small boat buried in the banks of the river mouth that empties into that bay, investigations were completed both onshore

and offshore to determine potential material remains. Two shovel tests were dug along the embankment, but no remains were found. In 2022, subsequent investigations with metal detectors were completed, but because of the high levels of ilmenite sands, this method was also unsuccessful. The visibility is higher in this area of the bay due to the high energy resulting in low turbidity and a hard bottom.

The first snorkel survey in Little Turtle Bay resulted in the discovery of a shipwreck. The remains are neatly scattered along the seafloor in a distinguishing pattern that speaks to how the wreck likely occurred. It appears that the vessel hit the rocky outcropping jutting into the bay and the remains were scattered almost parallel to the shore, likely carried by the localized currents. The wreck lies in roughly 4.5 m of water and is fairly accessible by snorkel, but SCUBA was necessary to map and record the wreck scatter. During the initial setup for mapping, the team was blown off the wreck and forced to abandon efforts to fully record the vessel. However, ropes coiled and hardened by concretion were determined to be likely 20th century in origin. Copper sheathing, the ship's gudgeon, and a piece of copper pipe fitting were also found. This evidence leads to the conclusion that this is potentially a 20th-century steam-powered vessel. Strong weather patterns prevented further investigation during the 2022 field season at this site, but plans to conduct further investigation during the 2023 spring field season are in place, when weather conditions are expected to be calmer.

Little Turtle Bay became the emphasis during the 2019 field season, where a few transects were conducted in Woodford Hill A. Survey was conducted near the extant bollard and, though visibility was limited, noticeable ballast was identified in the basin. Evidence of wreckage was noted from copper nails washed up on the rocky outcrops that connect the exposed rock with the bollard and beach. These findings led to the need for more research and surveying to be conducted in the bay. Unfortunately, due to the COVID-19 pandemic, this was not possible until 2022.

Field Season 2022

The 2022 field season consisted of several different methodologies to investigate the study area including geoarchaeological and maritime archaeological methods. The geoarchaeological goal of this field season was to find an environmental baseline for the area pre-contact, and hopefully, prior to human occupation and to determine whether the basin of the bay is an area of deposition. The maritime archaeological goals were to complete a systematic survey of the bay, documenting sites and materials related to human activity and any anthropogenic features of the bay.

Methodology

A systematic bathymetric map of the bay has not been previously conducted and the few maps that contain depth soundings for the bay are rough estimations at best. A Garmin Echomap UHD and down-scan transducer were used to gather depth data to create a more accurate model of the bathymetry of the anchorage. The down-scan sonar was not an ideal instrument, but due to unforeseen circumstances, planned equipment did not make it down to the island and the team improvised with locally available equipment. The Garmin Echosounder was not a self-contained unit, so a custom portable system had to be built using a car battery, gator clips, a toolbox, hose clamps, a C-clamp, PVC pipe, wood scraps, zip-ties, and duct tape.

A local fishing boat ran transects in the bay, but due to hidden rocky outcrops and difficult conditions the transects were not straight lines parallel to shore. Gaps in coverage parallel to the coast were filled in using a kayak to record depths and look for anomalies in the shallow waters and in and around reef heads. In total, four limited surveys were completed, two by boat and two by kayak.

Garmin no longer has the software to extract the data from the echosounder, so a trial of Sonar TRX software was used to extract the depth, the map of the transects, and the coordinates of points-of-interest. The comma-separated values (CSV) files were then processed in ArcGIS Pro in both 2D- and 3D-scenes. These data were combined with global positioning system (GPS) points recorded with a Garmin Mk2i Descent dive watch to incorporate depth data from the diver-led surveys. A 3D-elevation model was created from the collected data.

The bathymetric data gathered were essential to determining areas to core. The goal was to find basins that could potentially be sediment catchments to build an understanding of how and what sediments are deposited in the bay and at what rate. Poorly circulated coastal environments, such as semi-enclosed lagoons, are excellent sediment archives because these habitats preserve high-resolution records of natural and anthropogenic events. Sediment cores collected from comparable settings in other Caribbean islands have exhibited well-preserved millimeter (mm)- to centimeter (cm)-scale laminations (sequences of fine layers) representing environmental changes that can be tied to specific natural or anthropogenic events (Brooks et al. 2007; Larson et al. 2015; Brooks et al. 2015). High-resolution, short-lived radioisotope (SLRad) age dating of cores will be used to correlate natural (e.g., volcanic eruptions) or anthropogenic (e.g., clear cutting) events with sedimentological signatures. In summary, the sedimentological signatures of a down core will create a time sequence, which will be used to reconstruct the natural and cultural evolution of LaSoye Bay. Two cores were collected from a depth of 50 feet (ft.) in the center of one of the two basins identified with the FishFinder. Both were aluminum cores taken by divers by hand. Grab samples were collected in cardinal directions around the core. A third marine push-core was collected in shallow water. This site was chosen for a core because of a change in the sediment composition in the 10–12 ft. range, resulting in a much coarser-grained sediment, with more shell and rocks present.

The cores will be used to determine areas of deposition and create a baseline for large-scale climatic events taking place within the bay. Through sedimentology, it can be determined whether these sediments are terrgerious (the result of erosion), or if they are primarily marine sediments. Shifts between the two can help ascertain storm activity in the environmental record, which can be dated using pb210 (Larson et al. 2015).

Due to low visibility, the survey methodology was targeted based upon areas of potential interest flagged using the FishFinder. Areas of interest (AOI) had to be investigated slowly, and close to the bottom due to the high levels of sedimentation suspended in the water column.

A pilot study began in 2021 to map in the shoreline yearly using an Emlid RTK GPS (cm accuracy) to map in the shoreline and establish potential rate of loss year-over-year. Little rate of change has been noted over the past year. The lack of major storms over this period also lends itself to this result. More data will be collected in 2023 to add to this developing model.

Results

Figure 2 shows the 2D-model of the map created from the Garmin UHD FishFinder CSV files and the depth data from the Garmin Mk2i Descent dive watch. The map still has some major gaps, due in part to high coral reliefs and rock outcrops. These data offer insight into how the bay was used and manipulated to create an anchorage and an active harbor for both shipping and fishing.

A clear 100+ m wide pathway into the bay is framed by large rocky outcrops. This path is aligned with the settlement found on Woodford Hill beach. This was verified by divers and found to be clear of obstructions, with a sandy bottom and surprisingly clear waters. The path then opens into a small, deep basin in front of Woodford Hill A, parallel to the abandoned lime works located on the neighboring beach.

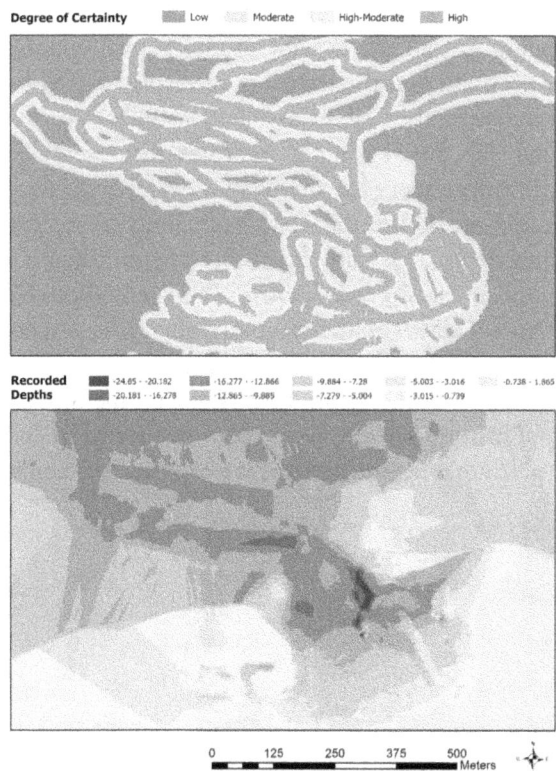

FIGURE 2. Results from the bathymetric map. (Figure created by the authors, 2022.)

Lab activities are being carried out at the Eckerd College Marine Geology Laboratory. Sediment samples are being analyzed for texture and composition to isolate sedimentological signatures. Samples are being dated using short-lived radioisotopes (SLRad) at sub-decadal scale to correlate the sedimentary signatures to specific depositional events, which can then be tied to either anthropogenic or natural catastrophic processes responsible for each depositional event. Preliminary results from the cores signify that the bay is a catchment zone and accumulates sediments at a high rate indicating poor circulation, which is interesting given the high-energy environment. Signatures down core are correlated to past hurricane events.

From the initial bathymetric data and diver ID, it appears that ships would need to enter from the northeast of the bay and hook southwest to safely enter the channel. The tops of some of the adjacent rock outcrops are so shallow to prohibit even divers from passing over them. These would have proved to be extremely hazardous for ship navigation. It is along this rock edge that a variety of artifacts related to the long-term use of the bay were identified. Debris in the anchorage speaks to the long-term use of the area. Some of which are shown here, in Figure 3. This includes a buckle, 72 × 50 cm dating from the 1840s to 1860s, likely used with a women's fabric belt (DAACS 2022). Unfortunately, during cleaning and prepping to put the buckle through electrolysis, it proved too fragile and began to crumble. Therefore, exact final measurements of the buckle after cleaning cannot be determined. Curved metal features measuring 1.5 m in diameter resemble barrel straps partially buried in the sediments near the channel, but could be related to other industrial undertakings in the area. Highly concreted pieces appear to be either part of a mast or part of an engine structure, which is roughly 1.2 m in length. Other unidentified objects and debris were recorded by divers to indicate long-term use of the area. Piles of bottles, some dating to the 1930s, were found scattered around the bay, but tend to collect in larger piles around outer reefs, likely being dragged out into the deeper areas of the bay by the currents and catching on the reef structures. These deeper areas are catchment zones, with deep sediment which also catches material culture.

Woodford Hill Wreck

In 4 m of water, a shipwreck was discovered. The remains are lightly buried in sediment directly off Woodford Hill Beach A. Hand fanning was utilized to uncover the site and determine the size of the remains. Base lines were set up on a north-south and east-west axis. The wreck lies on a northeast/southwest line, perpendicular to the shore. Six long planks are present that measure 12.5 m, 14.0 m, 14.5 m, 15.0 m, 13.5 m, and 6.5 m. These boards are 20.0-cm wide and 6.0 cm thick. Conditions were difficult to work in due to heavy surge. At the southwest end of the wreck, remains include three exposed bricks. These bricks are 20 cm long and 10 cm thick, with a fourth brick located on the side that was unable to be measured due to being deeply buried in sediment. The fourth brick was a white to yellowish color compared to the reddish orange bricks found at the southwest end of the wreck. These bricks are not machine made, and have smooth exteriors and visible inclusions. They are also longer and thicker than common machine-made bricks.

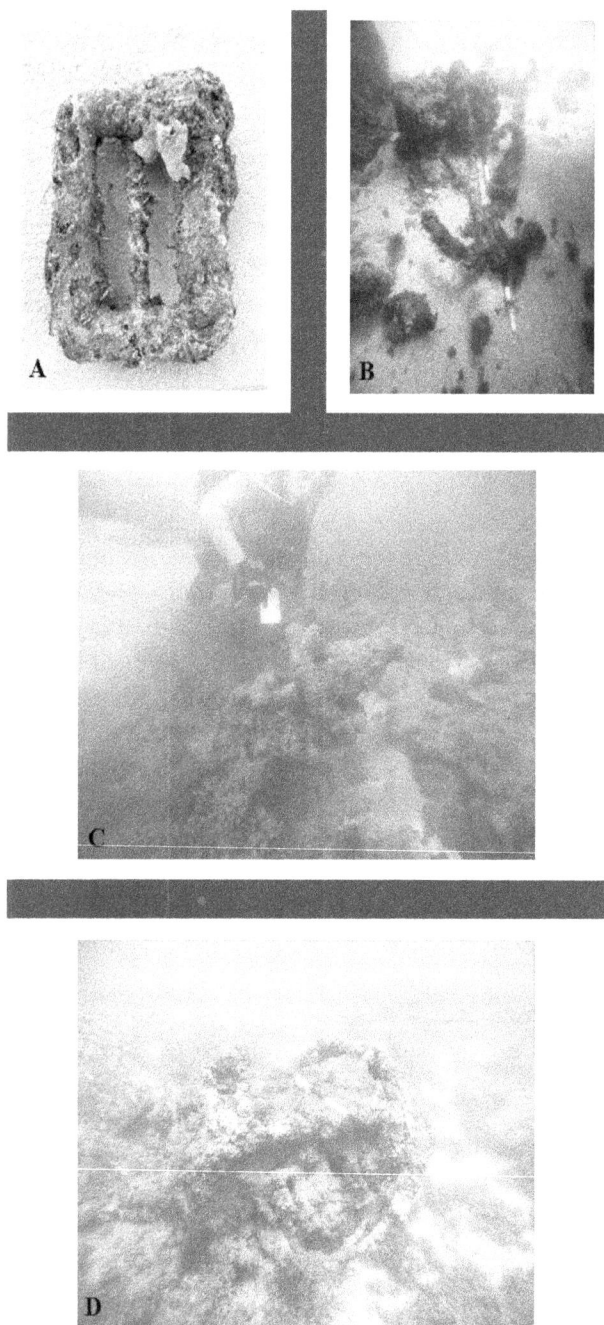

FIGURE 3. Images from the 2023 diver survey: A: Belt Buckle, B: Anchor, C: Possible Mast Fragment, D: UID Shipwreck Debris. (Photos by the primary author, 2022.)

Other interesting features documented during the brief survey was a chain and ring attached to the debris, as well as a concretion around the fastenings and coating parts of the exposed planks. The concretion has a slight blue-green tinge, which could indicate the use of copper sheathing. Approximately, 3 m from the wreck, a roughly 1 m piece of a chain was found. Given the frequent use of this area, it is difficult to determine if this is associated with the wreck but remains a possibility. Modern trash and debris, as well as a large piece of what appears to be modern farm equipment is nearby as well. Several modern objects, such as a spool for thread, were found wedged into the planks, indicating high energy and continued erosion from the shore.

The remains are incomplete and ample evidence of salvaging is present. Several of the planks appear to have been cut and the fasteners and nails have all been removed from the visible remains. The lack of iron oxide around the fastener holes indicates they were likely made from copper or a copper alloy, which was salvaged from the site. Given the wreck's proximity to shore and shallow depth, it is fairly accessible by snorkeling if the location of the site was known. The cut timber visible in Figure 4 shows evidence of salvaging. It is suspected that it took place soon after the wreck event as it not remembered by the community. Through the collection of oral histories, it has never been mentioned by locals and given that it is buried, the wreck is likely unknown.

FIGURE 4. From southwest: exposed plank thickness and evidence of salvaging, Woodford Hill Wreck. (Photo by the primary author, 2022.)

A dendrochronology sample was taken from a plank lacking knots and, in an area where the wood was intact, with little visible damage from microorganisms. The sample was sent to Alden Identification Services, where it was sorted based on microscopic characteristics via Minnesota State University's Inside Wood Database and to Thünen Institute of Wood Research in Hamburg for verification and identification using their wood collection of over 50,000 microscopic samples. Based on both analyses, overlapping agreement with the family Bignoniaceae was noted. The Tabeuia family has often been cited for ship construction, such as the *Tabeuia Rosea*, which offered a high level of agreement and the *Tabeuia heterophylla* (roble blanco), which is widely distributed throughout the Lesser Antilles and used for boat building (Meniketti 2012). Bignoniaceae family Tabeuia, which contains white cedar, is a wood used in the Caribbean for the construction of schooner barges, such as the Doughier and the Nevis lighter. The vessel measures at least 49 ft. long based on the remains and given the thickness of the planks and the presence of only one maststep, which is shifted forward on the vessel, it is highly probable that the vessel is a coastal barge.

Future Plans

Mapping and recording the Woodford Hill wreck will be a priority for the 2023 field season. A grid will be laid, and the area around the wreck will be ground-truthed and surveyed to determine the extent of the wreckage. Detailed maps, photographs, and videos will be obtained to create renderings, help identify potential boat typologies, and be used for public outreach. Hand fanning and minimal excavation with trowels has been approved by the Dominica Fisheries Division and will only be conducted to uncover identifying objects and/or pieces of the wreck to gain an understanding of the timeframe, use, and origin of the wreck. More dendrochronology samples will be taken to determine if the wreck was patched or if it is all built with one kind of wood. Archival research on the island will help shed light on activities taking place in the immediate area and will be combed for mentions of maritime-related activities with a thorough search of newspapers from the 18th and 19th centuries.

Interviews with local stakeholders and community members will be conducted to further bolster the story of activity in the bay into modern times and connect to the more recent history and living memory of the area. Informal interviews with fishermen and residents have already been insurmountably helpful in reconstructing past activity and understanding activity and movement within the area. The method used will be snowball sampling, beginning with established local connections, such as the fishermen in Calibishie local business

owners, and residents that have grown up around this stretch of coastline.

Another area to investigate is the new seaweed cooperative taking place in the bay. This began in 2022 and is currently a small operation and is being grown meters from the 2022 shipwreck site. This was started by one of the Calibishie village council members and this individual is another person who will be interviewed during the 2023 field season. This offers a fascinating insight into how community members continue to utilize the bay to provide sustenance and could potentially lead to another export industry (Neish et al. 2017). Seaweed farming has taken off in places such as Placencia, Belize (Carabantes 2020) where the focus has been on women seaweed farmers. It is used in local health drinks and shakes as well as being exported for use in beauty products.

Conclusion

Through two field seasons of diver-led surveys and three seasons of fieldwork, two shipwrecks have been discovered along with other artifacts and debris related to the long-term use at LaSoye Bay, Dominica. As a part of these surveys, the first rudimentary bathymetric model of the bay has been created.

Acknowledgements

This work would not have been possible without the assistance of Emily Schwalbe during the first field season. Eckerd College Department of Marine Geology and Rebkka Larson, Gregg Brooks, Vince Shonka, and Casandra Guzman helped facilitate the field and lab portions of the sediment analysis for the environmental interpretations. Oneida LG2 Environmental Solutions lent equipment and personnel, which made this project possible. Christine and Bradford Diehl for their generous hospitality and continuous support of this project. Research was funded in part by the PaleoWest Foundation and the University of South Florida Strong Coasts fellowship.

References

BOOMERT, ARIE
2009   Searching for Cayo in Dominica. Presented at XXIII Congress of the International Association for Caribbean Archaeology, Antigua, West Indies.

2011   Aspects of Island Carib Archaeology. In *Communities in Contact: Essays in Archaeology, Ethnohistory and Ethnography of the Amerindian Circum-Caribbean.* Corinne L. Hofman and Anne Van Duijvenbode, editors, pp. 291–306. Sidestone Press, Leiden, NL.

BOROMÉ, JOSEPH A.
1967   The French and Dominica, 1699–1763. *Jamaican Historical Review* 7(1):9.

BOWEN, THOMAS
1790   *A Map of the Island of Dominica Taken from an Actual Survey, also Part of Martinico & Guadalupe Shewing their True Bearing & Distance from Each Other.* London: Printed by E. Cave, 1778.

BROOKS, GREGG R., BARRY DEVINE, REBEKKA A. LARSON, AND BRYAN P. ROOD
2007   Sedimentary Development of Coral Bay, St. John, USVI: A Shift from Natural to Anthropogenic Influences. *Caribbean Journal of Science* 43(2):226–243.

BROOKS, GREGG R., REBEKKA A. LARSON, BARRY DEVINE, AND PATRICK T. SCHWING
2015   Annual to Millennial Record of Sediment Delivery to US Virgin Island Coastal Environments. *Holocene* 25(6):1015–1026.

CARABANTES, ESTENIA J. ORTIZ
2020   Using a Systems Thinking Approach and Health Risk Assessment to Analyze the Food-Energy-Water System Nexus of Seaweed Farming in Belize. Doctoral Dissertation, Department of Environmental Engineering, University of South Florida, Tampa, FL.

DAACS
2022 Object Query 3, August 1, 2022. The Digital Archaeological Archive of Comparative Slavery. < https://daacsrc.org/queries/query_objects_three_results>. Accessed 25 March 2023.

HAUSER, MARK W.
2011   Routes and Roots of Empire: Pots, Power, and Slavery in the 18th-Century British Caribbean. *American Anthropologist* 113(3):431–447.

HAUSER, MARK W, DOUGLAS V ARMSTRONG, DIANE WALLMAN, KENNETH G KELLY, AND LENNOX HONEYCHURCH
2019   Where Strangers Met: Evidence for Early Commerce at LaSoye Point, Dominica. *Antiquity* 27:1–8.

LARSON, REBEKKA A., GREGG R. BROOKS, BARRY DEVINE, PATRICK, R. SCHWING, CHARLES E. HOLMES, TOM JILBERT, AND GERT-JAN REICHART
2015   Elemental Signature of Terrigenous Sediment Runoff as Recorded in Coastal Salt Ponds: US Virgin Islands. *Applied Geochemistry* 63:573–585.

MENIKETTI, MARCO
2012   And Now They Are Gone: Documenting the Last Sailing Lighter of Nevis, West Indies. *The International Journal of Nautical Archaeology* 41(1):134–147.

NEISH, IAIN C., MIGUEL SEPULVEDA, ANICIA Q. HURTADO, AND ALAN T. CRITCHLEY
2017   Reflections on the Commercial Development of Eucheumatoid Seaweed Farming. In *Tropical Seaweed Farming Trends, Problems and Opportunities.* Developments in Applied Phycology, Vol 9, Anicia Hurtado, Alan Critchley, Iain Neish, editors, pp 1–222. Springer, Paradise, NS.

· · · · · · · · · · · · · · ·

Marie Meranda
Department of Anthropology
University of South Florida
4202 E. Fowler Avenue, SOC 107
Tampa, Florida 33620

Megan Bebee
Oneida LG2 Environmental Solutions
10475 Fortune Parkway, Suite 201
Jacksonville, Florida 32256

Does Evidence for Jewish Pirates Exist Archaeologically? A Case Study of Sinan Reis

Leah Tavasi

While piracy is a modern phenomenon as much as an ancient one, piratical theory is relatively opaque under the umbrella of maritime archaeology. Smugglers, buccaneers, and freebooter's fluidity and capriciousness are not reflected in the black-and-white morality of a quintessential pirate. After presenting a background in pirate theory, the historical and archaeological remnants of the Jewish pirate are investigated during the time of the Spanish Inquisition (1478–1834) through the case of Sinan Reis. Sinan, Hayreddin Barbarossa's captain and famous 16th-century Barbary corsair, is considered through surviving engravings, artwork, and artifacts that detail his voyages. Most compelling is a surviving engraving consisting of two Latin couplets that provide lively insight into Sinan's motivations as a pirate. In conjunction with pirate theory and known written history of Sinan, a picture of the life of a Jewish pirate is drawn. Conclusions highlight the use and power of pirate theory, the bias of contemporaneity in archaeology, and cautions that the application of the term "pirate" should be utilized with care.

Embora a pirataria seja um fenómeno tanto moderno como antigo, a teoria da pirataria é relativamente vaga no âmbito da arqueologia marítima. A fluidez e o capricho dos contrabandistas, bucaneiros e piratas não se refletem na moralidade a preto e branco de um pirata por excelência. Depois de apresentar os antecedentes da teoria dos piratas, são investigados os vestígios históricos e arqueológicos do pirata judaica durante o período da Inquisição espanhola (1478-1834) através do caso de Sinan Reis. Sinan, o capitão de Hayreddin Barbarossa e famoso corsário da Barbária do século XVI, é analisado através de gravuras, obras de arte e artefactos que sobreviveram e que detalham as suas viagens. O mais interessante é uma gravura que consiste em dois versos latinos que fornecem uma visão viva das motivações de Sinan como pirata. Em conjunto com a teoria dos piratas e a história escrita conhecida de Sinan, é traçada uma imagem da vida de um pirata judeu. As conclusões sublinham a utilização e o poder da teoria dos piratas, o preconceito da contemporaneidade na arqueologia e advertem para o facto de a aplicação do termo "pirata" dever ser utilizada com cuidado.

Alors que la piraterie est un phénomène moderne autant qu'ancien, la théorie piratique est relativement opaque sous l'égide de l'archéologie maritime. La fluidité et le caprice des contrebandiers, des boucaniers et des flibustiers ne se reflètent pas dans la moralité en noir et blanc d'un pirate par excellence. Après avoir présenté un fond dans la théorie du pirate, les restes historiques et archéologiques du pirate juif sont étudiés à l'époque de l'Inquisition espagnole (1478-1834) à travers le cas de Sinan Reis. Sinan, le capitaine de Hayreddin Barbarossa et célèbre corsaire barbare du 16e siècle, est considéré à travers des gravures, des œuvres d'art et des artefacts survivants qui détaillent ses voyages. Le plus convaincant est une gravure survivante composée de deux couplets latins qui fournissent un aperçu vivant des motivations de Sinan en tant que pirate. En conjonction avec la théorie du pirate et l'histoire écrite connue de Sinan, une image de la vie d'un pirate juif est dessinée. Les conclusions soulignent l'utilisation et la puissance de la théorie du pirate, le biais de la contemporanéité en archéologie, et avertit que l'application du terme « pirate » doit être utilisée avec précaution.

Introduction

If one was to ask scholars of today fluent in piratical theory to define its very nature, I am confident their first words would center upon a message of opacity. The black and white morality tied to the words "pirate" or "piracy" do not reflect the fluidity and capriciousness these roles filled throughout history. Yet modern popular culture, courtesy of the animators at Disney, brings to mind a distinct initial image: a rough and rowdy figure with a flawed moral compass. However, the realities of piracy should be much less foreign than it is today, for piracy is as much a modern concept as it is ancient. Tales of the Sea Peoples from ancient Egypt or Pompey's "triumph" over Cilician pirates have been tied in motivation to pirates of the modern era, from the Barbary corsairs, escapades in the New World, and more recently the Somali pirates.[1] I attempt to do the same with a niche sect of piratical behavior: the seldom explored Jewish pirate. The introduction of this article first and foremost serves as a discussion of essential theory: what it means to "be" a pirate, what motivates piracy, and how pirates are recognized both historically and archaeologically. These fundamentals are necessary for further study into Jewish piracy and its archaeological sources. A discussion of historical and archaeological evidence, in the form of surviving engravings, cameos, and artwork, of Sinan Reis will follow using the aforementioned outlined theory as a guide. The results from this case study are then used to draw conclusions on the potential and function of piratical archaeology and pirate theory.

Definitions and Pirate Theory

What is a pirate? As alluded to above, many shades of gray exist in the definition of piracy: between pirate and privateer, smuggling, freebooting, buccaneering, and so on. Daniel Defoe's *A General History of the Pyrates* (1724) defines piracy by those who committed economical and logistical feats against the state. Pirates are, in the most basic terms, sailors who seize property by force (Dawdy and Bonni 2012). The actions of pirates and privateers are today, as they were in the past, very similar, with the only difference between them being a *patente*, or letters of marque, from a sanctioned state (Dawdy and Bonni 2012; Skowronek 2016). Privateers also seize property; they simply have legal license to do so. Both roles involved the same actions of violence, and for the purposes of this essay these similarities and subtle differences between pirate and privateer are highlighted. For one country's pirate is another country's hero; the label of a pirate is often misleading. As will be shown, it is the surviving Christian narrative that delineates pirate versus war hero.

It is the onus of the victor to determine who is "right"—a decision written more in stone than sand—often remembered without the nuances that color reality. Skowronek and Ewen (2006: 261) compare the sentimentalities of English "heroes" such as Sir Frances Drake and Sir John Hawkins between the English and Spanish; to the Spanish, they were regarded "very much the same as Americans [in 2006] regard[ed] Osama bin Laden and Saddam Hussein." This disparity in opinion is monumental. Moreover, within a country, the "enemy" can change so flippantly that someone acting as a privateer could be later condemned a pirate, such as in the cases of William Kidd and Henry Morgan. William Kidd, flying only British colors (Hatch 2016),

was executed for piracy for attacking French ships after a change in Britain's political climate (Zacks 2002). Henry Morgan, on the other hand, was shipped and imprisoned in the Tower of London. Four years later, he was elevated by King Charles II as Sir Henry Morgan; now knighted, and in the position of lieutenant governor of Jamaica (Forbes 1948), his first task was to squash the issue of piracy (Kritzler 2008). In other words, Henry Morgan was tasked with putting an end to the very actions that brought him his success. Privateers were well versed in the dubious legality of their actions, allowed only by the permission of the crown (Benton 2005), and this suggests not only a disconnect between European history's representations of the black-and-white line of lawlessness, but also a moral compass on the side of the pirate.

Furthermore, archaeological theorist Vincent Gabrielson (2013) shows sufficient evidence that piracy's cradle can be situated at the very top of the political and socio-economic ladder. Using Herodotus' *Histories*, he shows how two prominent political figures in the late 6th and early 5th-centuries BC, Histiaios of Miletus and Dionysios of Phokaia, were practiced systemic sea raiders (Herodotus, *Histories* 6.5–6.17). Dionysios of Phokaia specifically did not attack Greeks (Herodotus, *Histories* 6.17), demonstrating an ethnic bias for his violence (Gabrielsen 2013). With these two historical figures, Gabrielsen demonstrates a lack of class distinction within piratic behavior; pirates could be both community leaders and renegades, using both avenues to fulfill an agenda. This concept of piracy—acting on behalf of a political player, for a political reason—is seen in the evidence of Jewish piracy. Governments and politically concerned private citizens biased themselves against the Spanish and their Holy War, choosing to ally themselves with powers more tolerant of religious freedom.

The next logical step to understand pirate theory is to look at a pirate from an anthropological perspective. What leads a group of people to piracy? What are their motivations? Archaeological theorists, such as Aaron Beek (2015), have analyzed the classical world, predominantly Pompey's tirade against piracy, to determine the "push" and "pull" mechanisms that underlie such motives. Quoting Cassius Dio: "feed your slaves, so that they may not turn to brigandage" (Cassius Dio, *Historiae Romanae* 77.10), Beek intonates through his continual motivation approach that the problem is systemic. "Push mechanisms," constitute hunger, poverty, or social strife (Gabrielsen 2013); people that are hungry or poor will be pushed towards piracy. Unsurprisingly, peaceful times, such as during the *Pax Romana* (27 BC–180 AD), saw lower levels of piracy than those which were more desperate (Beek 2015). People resort to piracy because the system itself is broken, whether for survival or for retribution. To foreshadow comparisons to come, against the backdrop of the Spanish Inquisition (1478–1834), Jewish piracy will be shown to be fueled by the repetitive and persistent Spanish violence and persecution towards the Jewish people.

So far, an understanding of the definitions, motivations, and biases of piracy have been discussed. It is with this basis in theory that the archaeology can then be analyzed: how would an archaeology of piracy present itself? As underdeveloped as is piratical theory, even more so is the archaeology of piracy. What does a pirate ship look like? How does one tell the difference between a pirate ship and a merchant ship or a war ship? Would there be evidence of lairs, graves, artillery, evidence of smuggling and contraband? Did pirates dress differently? What textiles would one look for? The archaeology of piracy is a facet that, until the last decade, was relatively unexplored.

Tackling the problem in groundbreaking research, recent scholarship has begun to explicitly state where the physical evidence of piracy would exist (Ewen 2006; Skowronek 2016). Lairs, or the staging sites and bases of piracy, are the first category that would show evidence of illicit activity. Secondly, pirate ships, if able to be correctly identified, would be a clear indicator of piracy. Thirdly, the remnants of the actors or pirates themselves, any material evidence they would leave behind, would lend itself to piracy. With this combination of underwater and terrestrial sites, an outline of a methodology is constructed.

None of these are easily accomplished. Famous lairs that have been exposed and explored include Ile de Tortue (Tortuga) and Ile Sainte Marie, both of which served as a pirate base in the 17th and 18th centuries. Specific to Jewish pirates, a potential lair location highlighted in this paper is the island of Jerba, and is drawn from historical references of activity. It is important to recognize that archaeologists are, in essence, looking for places that were meant to be hidden. Pirate ships are even more elusive. Some pirate ships have been successfully identified and excavated through a combination of negative evidence and the historical record, such as Blackbeard's vessel, the *Queen Anne's Revenge*, off the coast of North Carolina (Wilde-Ramsing and Ewen 2012). Unfortunately to date, Jewish pirate ships such as the *Queen Ester* or *Prophet Samuel* (Tolkowsky 1964)

exist only in the historical record. Lastly are the artifacts of pirates, pirate victims, and material culture of smuggling. Henry Morgan, a pirate and privateer that became Jamaica's lieutenant governor in 1678, is clearly attached to Port Royal. However, material evidence attached to him, and the island, is limited. A record of objects bequeathed to his wife after his death exists, with an obvious show of expensive goods (Ewen 2006); however, at this point he was an established and legal citizen of the English government. Can that be considered pirate bounty? The above examples come explicitly from 18th and 19th-century instances of well-documented piratic behavior; looking for evidence of Jewish piracy focusing on material evidence is therefore, understandably, not easy.

Lastly, defining a Jew, much like defining a pirate, during the colonization of the Americas can be difficult. Throughout the duration of the Spanish Inquisition, Jews were as forbidden in the New World as in the Old. Whether one was a true convert and believer in Christ, an atheist, or a covertly practicing Jew, all were considered illegal. They were colloquially referred to in dispatches and letters of the time as *Conversos*, translating to converters: "New Christians" who converted to Christianity during the Inquisition rather than face the consequences from the High Inquisitor (Wiznitzer 1960). The terms Portugals, or *Marranos*, translating to pigs, were also used (Tolkowsky 1964). Many *Conversos* secretly maintained their Jewish heritage and were, therefore, under constant scrutiny from the High Inquisitor. These "Holy Wars" spanned centuries and led to massive shifts in alliances, secrecy, and in population from continent to continent.

This section has served to define piracy in archaeological and anthropological contexts. This paper continues through a brief exploration of the remnants of Jewish piracy through the historical record, which are then analyzed against modern pirate theory with reference to material culture. First, Sinan Reis, the "most famous Jewish pirate" of the 16th century, is explored by looking at surviving engravings, using them in conjunction with the historical record to determine the motivations behind his attacks.

Sinan, the "Famous" Jewish Pirate

Sinan, also referred to as Sinan Coefut, Sinan Reis, and later Kaptan Sinan Pasha, was a Barbary corsair who served the Ottoman Sultan Suleiman during the 16th century. From Smyrna, modern day Izmir, Sinan, a Jew (Figure 1), became a trusted captain of Hayreddin

Barbarossa, one of the most famous Barbary corsairs (Fuchs 2000; Tolkowsky 1964), and eventually was promoted to report directly to the Sultan. Barbarossa came from the Greek island of Mytilene, now the capital city of Lesbos. Sinan's connection to Barbarossa exists in physical evidence today through artwork, as Sinan is typically depicted at Barbarossa's side (Figure 2). A lapis lazuli broach sold by Sotheby's in 2012 depicts Sinan and Barbarosa back-to-back (Figure 3). Both hailing from the Ottoman Empire, which was relatively less biased against Jews in the 16th century (Lybyer 1913), Sinan was able to rise up in the ranks of the Ottoman Empire (Karzek 2021; Malcolm 2016). His family was a victim of forced expulsion from Spain during the Inquisition (Tinniswood 2011). I argue this was the beginning of the aforementioned ethnic bias that served him throughout his career. Protection from the Sultan gave him legal grounds to take his revenge upon those who forced his family out of Spain (Tolkowsky 1964). Material evidence of his personal allegiance and expulsion is found in a surviving engraving (Figure 4). However, it is unclear whether his Ottoman allegiance or his desire for revenge against the Spanish motivated these attacks. This 16th-century engraving sheds light on his potential motivations with its descriptive Latin couplets (Figure 4). The medieval Latin, edited in brackets to reflect classical declensions and conjugations has been included in the figure caption.

In this engraving, Sinan is proudly self-identifying as a pirate, as well as paying homage to his Jewish heritage. He suggests that it was his involuntary removal from his homeland that forced him into piracy. Spanish Jews who refused to be baptized were forced to find new homes; many of them traveled to North Africa and the eastern Mediterranean (Tolkowsky 1964). The second couplet is more personal. In 1539, Sinan's son was kidnapped by the Lord of Elba, after Charles V recaptured Tunis and took Algiers from Sultan Suleiman. It took Sinan and Barbarossa five years, until 1544, to successfully sack Piombino and rescue his son (Tolkowsky 1964). As established in the introduction, "push" and "pull" mechanisms prompt piracy. Mandatory deportation and the kidnapping of a child would certainly spur ill feelings towards an empire. In the historical sources that describe Sinan's career and Barbarossa's, he launched an attack against the Spanish at least 17 times (Tolkowsky 1964; Lybyer 1913; Karzek 2021; Malcolm 2016). From this lens, it is clear that he allied himself with the Ottoman Empire, which gave him the opportunity to attack the Spanish.

What is known of his moral compass? Sinan is noted in the historical record for his humane treatment of prisoners. When Charles V was close to recapturing Tunis from the Ottomans in 1534, Barbarossa wanted to slaughter the 20,000 Christian slaves in the city's underground dungeons. Sinan dissuaded him, saying "to stain [our]selves with so awful a massacre would place [us] outside the pale of humanity forever" (Currey 1910:108). In another instance, Sinan, after becoming the governor of Algiers in 1551, commanded the Turkish navy. He captured Tripoli, and with it, the very knights of Malta that bested him in their recapture of Tunis in 1534. Gracious in victory, in contrast to Charles' massacre after his recapturing of Tunis, he let the knights go. With regards to King Charles' capture of the city in 1534, quotes from primary sources describe the event as "the day of the Lord's wrath" (Hirschberg 1974: 480;

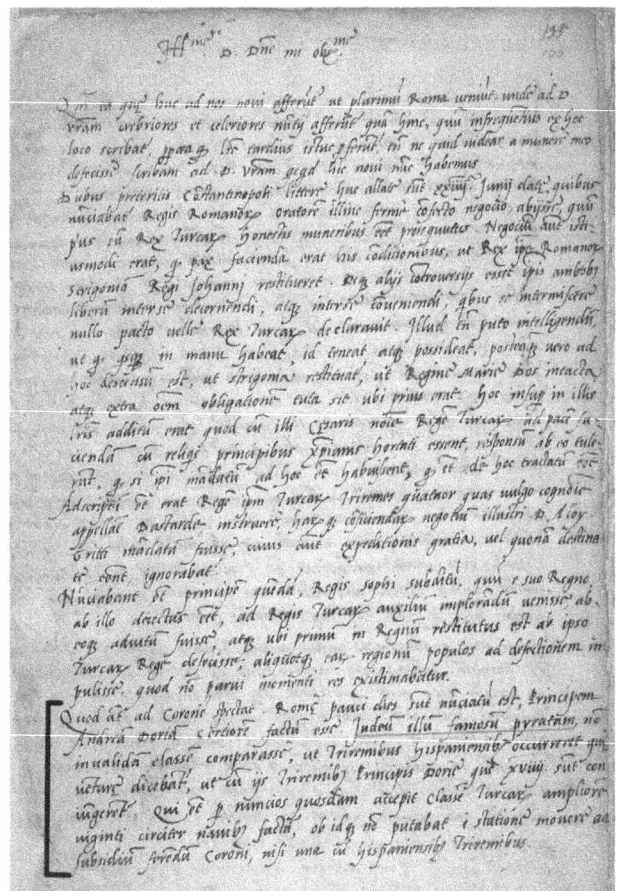

FIGURE 1. The first page of a letter from John Casali, the English Ambassador in Venice, to the Duke of Norfolk, written on 16 August 1533. The markings on the left-hand side highlights a paragraph where he writes of Judeu[m] illu[m] famosu[m] pyratam or "the famous Jewish pirate." (© British Library Board [Cotton MS Nero B VI 62 folio 135].)

FIGURE 2. Material evidence of Sinan and his role as a pirate of the corsairs. A tempera on canvas of Sinan (d.1546) and Barbarossa (1478–1546), from left to right. Painted in 1535, by an imitator of the Gentile Bellini. (Figure provided with permission from the Art Institute of Chicago via Creative Commons Zero program.)

FIGURE 3. Material evidence of Sinan and his role as a pirate of the corsairs. A 16th-century carved cameo made of lapis-lazuli, silver, and agate: from Lot 123 of auction L12225 held on April 24th, 2012 in London. Depicting Barbarossa and Sinan back-to-back: Barbarossa facing right, Sinan facing left. (Figure provided with permission from Sotheby's.)

Hanna 2015; Cardozo 1940) and that all were "smitten with the edge of the sword when the uncircumcised came to the city" (Hirschberg 1974: 480). This lends to a much more realistic portrayal of a pirate, one that is not entirely morally corrupt, but rather who participated in the same bouts of violent activity and acts of benevolence common to a state.

Some of Sinan's last recorded actions are his pirate raids in the 1550s that he conducted from the island of Jerba (modern day Djerba off the Tunisian coast) on the coastal districts of southern Italy and Sicily (Brewer 1862; Ormerod 1924). A terrestrial and water survey in pursuit of post medieval archaeology could provide evidence of such raids. Studies in the area have been conducted in terms of classical Roman archaeology (Fentress et al. 2009), such as research into the island's aqueducts and its connection to the Murex dye trade. Its early, 4th- to 5th- century Before Common Era (BCE) Jewish communities have also been studied (Udovitch and Valensi 1984; Fentress et al. 2009). No late medieval or post-medieval archaeological survey in the area to date has occurred, which highlights the bias of location and contemporaneity in archaeology. Following the outline given by Skowronek and Ewen (2006) in their treatise on pirate archaeology in the Americas, targeting this area to discover a lair dating to the time of Sinan and Barbarossa would be the next logical step.

Conclusion

The archaeology of piracy is difficult to pinpoint, and even more so is the archaeology of Jewish piracy. It is made challenging by the logistics of piracy, such as shipwreck distribution bias, or attempting to find lairs and smuggling operations people were actively trying to hide. It is also problematic because the label "pirate" is, in most cases, an applied identifier. In the absence of historical documentation, or someone's bona fide attestation that they considered themselves a pirate, how does one prove that someone is a pirate? Pirates were multifaceted and existed on the spectrum from leftist anarchists to aggressive capitalists, rebels and sinners, radical thrifty tradesmen, and spenders with excessive vices. They filled no specific niche. It is also equally possible that figures of the past that are labeled as pirates today did not consider themselves pirates at all. Perhaps, for example, an alternative depiction of Sinan, the famous Jewish pirate, is the "correct" one; perhaps Sinan viewed himself as a loyal captain to his Empire and to the Suleiman, and it is on the account of the engraver of the print (Figure 4) and the historical documentation of the Christian elite (Figure 1) that he is considered a pirate today. Does his depiction in photographs alongside Barbarossa, a documented sea raider, confirm his own profession? Not necessarily; it is a concept certainly worth consideration.

Nevertheless, from this research, it is clear that a more in-depth foray and analysis into Jewish piracy is necessary to tease out such nuances. Field surveys can be conducted in the targeted areas discussed; Djerba would make an excellent starting point for further data

collection. What is indisputable is that piracy and privateering, whatever the label, was an art form of expansion (Ormerod 1924) characteristic during the Age of Discovery. Pirate archaeology is a field with a renewed sense of purpose within the realm of the maritime cultural landscape. It acts as an important avenue of interpretation and can guide field research in addition to a more authentic illustration of piracy of the past. Using it, material evidence of even the most persecuted groups exists, if one is looking.

Acknowledgments

I would like to thank the University of Oxford for the access to research materials and Damian Robinson for his guidance and support. I would also like to thank Daniel Whittle, who lent his expertise in Latin translation.

Endnotes

1. *Roth (2020); Woodward (2004); Ganser (2020) are among many references bridging pirate theory through time, which the author suggests for further reading.*

2. *I argue here that Smyrna is the subject of both the first and second phrase of the couplet.*

FIGURE 4. An engraving by Theodoor de Bry of Sinan printed in 1596 as part of the series *Vitae et icones Sultanorum Turcicorum*. It is written in the meter of two elegiac couplets. The text roughly translates to, "Smyrna is my homeland, my parents are from the Jewish people. Smyrna[2] made me into a pirate on traveled seas. While I held my son, who had been brought back to me, my arms which desired him. A sudden death took me away with its wild joy."

> *Smyrna mihi patria est Ivdaea ex ge[n]te[s] pare[n]tes*
> *pyratam emensi me tvlit vnd[e] sali,*
> *dvm redvcem optatis complexib[us] implico natvm*
> *me svbita extinxit more fera laetitia.*

(Figure provided with permission from the Heidelberg University Library.)

References

BEEK, AARON L.
2015   Where have all the pirates gone? In *Invisible Cultures: Historical and Archaeological Perspectives*, Viola Gheller and Francesco Carrer, editors, pp. 270-284. Cambridge Scholars Publishing, Newcastle upon Tyne.

BENTON, LAUREN
2005   Legal Spaces of Empire: Piracy and the Origins of Ocean Regionalism. **Comparative Studies in Society and History** 47(4):700–724.

BREWER, JOHN. S. (EDITOR)
1862   *Letters and Papers, Foreign and Domestic, of the Reign of Henry VIII: Preserved in the Public Record Office the British Museum, and elsewhere in England.* Longman, Roberts & Green, London.

CARDOZO, MANOEL S.
1940   The Collection of the Fifths in Brazil, 1695-1709. *The Hispanic American Historical Review* 20(3):359–379.

CURREY, E. HAMILTON
1910   *Sea-wolves of the Mediterranean the Grand Period of the Moslem Corsairs, with Portraits and Illustrations.* Dutton, New York.

DAWDY, SHANNON L., AND JOE BONNI
2012   Towards a General Theory of Piracy. *Anthropological Quarterly* 85(3):673–699.

DEFOE, DANIEL
1724   *A General History of the Robberies and Murders of the Most Notorious Pyrates.* Ch. Rivington, London.

EWEN, CHARLES. R.
2006   Introduction. In *X Marks the Spot: The Archaeology of Piracy*, Russel K. Skowronek and Charles. R. Ewen, editors, pp. 1–12. University Press of Florida, Gainesville, Florida.

FENTRESS, ELIZABETH, A. DRINE, AND RENATA HOLOD
2009   *An Island through Time: Jerba Studies.* Journal of Roman Archaeology, Portsmouth, Rhode Island.

FORBES, ROSITA
1948   *Sir Henry Morgan: Pirate and Pioneer.* Cassell, London.

FUCHS, BARBARA
2000   Faithless Empires: Pirates, Renegades, and the English Nation. *ELH* 67(1):45-69.

GABRIELSEN, VINCENT
2013   Warfare, Statehood, and Piracy in the Greek World. In *Seeraub im Mittelmeerraum: Piraterie, Korsarentum und Maritime Gewalt von der Antike bis zur Neuzeit*, Nikolas Jaspert and Sebastian Kolditz, editors, pp. 133–153. Brill, Leiden.

GANSER, ALEXANDRA
2020   *Crisis and Legitimacy in Atlantic American Narratives of Piracy.* Springer Nature, London.

HANNA, MARK G.
2015   *Pirate Nests and the Rise of the British Empire, 1570-1740.* University of North Carolina Press, Chapel Hill.

HATCH, HEATHER
2016   Signaling Pirate Identity. In *Pieces of Eight: More Archaeology of Piracy*, Charles Ewen and Russell Skowronek, editors, pp. 208-227. University Press of Florida, Gainsville.

HIRSCHBERG, H. Z.
1974   *A History of the Jews in North Africa.* Brill, Leiden.

KARZEK, SAIM A.
2021   Ottoman Corsairs in the Central Mediterranean and the Slave Trade in the 16th Century. Master's Thesis, Department of History, Bilkent University. ProQuest Dissertations Publishing, Ann Arbor, MI.

KRITZLER, EDWARD
2008   *Jewish Pirates of the Caribbean.* Doubleday, New York.

LYBYER, ALBERT H.
1913   *The Government of the Ottoman Empire in the Time of Suleiman the Magnificent.* Harvard University Press, Cambridge.

MALCOLM, NOEL
2016   *Agents of Empire: Knights, Corsairs, Jesuits and Spies in the 16th-century Mediterranean World.* Penguin Books, London.

ORMEROD, HENRY A.
1924   *Piracy in the Ancient World: An Essay in Mediterranean History.* Liverpool University Press, Liverpool.

ROTH, ROMAN
2020   The Colonisation of Pontiae (313 BC), Piracy and the Nature of Rome's Maritime Expansion before the First Punic War. In *Piracy, Pillage and Plunder in Antiquity: Appropriation and the Ancient World*, Richard Evans and Martine De Marre, editors, pp. 84–96. Routledge, Abingdon.

SKOWRONEK, RUSSELL K.
2016   Setting a Course Towards the Archaeology of Piracy. In *Pieces of Eight: More Archaeology of Piracy*, Charles Ewen and Russell Skowronek, editors, pp. 1–14. University Press of Florida, Gainsville.

SKOWRONEK, RUSSELL K. AND CHARLES R. EWEN
2006   Identifying the Victims of Piracy in the Spanish Caribbean. In *X Marks the Spot: The Archaeology of Piracy*, Russel K. Skowronek and Charles. R. Ewen, editors, pp. 248–270. University Press of Florida, Gainesville, Florida.

TINNISWOOD, ADRIAN
2011   *Pirates of Barbary.* Vintage, London.

TOLKOWSKY, SAMUEL
1964   *They Took to the Sea.* T. Yoseloff, New York.

UDOVITCH, ABRAHAM. L., AND LUCETTE VALENSI
1984   *The Last Arab Jews: The Communities of Jerba, Tunisia.* Harwood Academic, New York.

WILDE-RAMSING, MARK U. AND CHARLES R. EWEN
2012   Beyond Reasonable Doubt: A Case for Queen Anne's Revenge. *Historical Archaeology* 46(2):110-133.

WIZNITZER, ARNOLD
1960   *Jews in Colonial Brazil.* Columbia University Press, New York.

WOODWARD, G. THOMAS
2004   The Costs of State–Sponsored Terrorism: The
        Example of the Barbary Pirates. *National Tax
        Journal* 57 (3):599–611.

ZACKS, RICHARD
2002   *The Pirate Hunter: The True Story of Captain Kidd.*
        Hyperion Books, New York.

· · · · · · · · · · · · · · · ·

Leah Tavasi
Institute of Archaeology
University of Oxford
36 Beaumont Street
Oxford, United Kingdom
OX1 2PG

Underwater Cultural Heritage Conservation and Climate Change in Canada

Aimie Néron

The United Nations Educational, Scientific and Cultural Organization launched the Decade of Ocean Science for Sustainable Development (2021–2030) that highlights the need for better collaborative approaches in marine sciences for ocean conservation and sustainability. Underwater archaeology is a diverse science that has been developing and participating in multidisciplinary projects. Given the impacts of climate change, submerged heritage in Canada is fragile and threatened, and remains scarcely documented. Measures are, therefore, necessary to protect knowledge on maritime and underwater heritage, both tangible and intangible, and understand the current and future impacts on the preservation and degradation of these unique non-renewable resources. The Underwater Archaeology Team projects from Parks Canada showcase the consideration and importance of the conservation of both natural and cultural heritage as underwater archaeology contributes to the understanding of humankind's relationship with the ocean and the now urgent need for adaptation to climate change.

A Organização das Nações Unidas para a Educação, a Ciência e a Cultura lançou a Década da Ciência dos Oceanos para o Desenvolvimento Sustentável (2021-2030) que destaca a necessidade de melhores abordagens de colaboração nas ciências marinhas para a conservação e sustentabilidade dos oceanos. A arqueologia subaquática é uma ciência diversificada que tem vindo a desenvolver-se e a participar em projetos multidisciplinares. Tendo em conta os impactos das alterações climáticas, o património submerso no Canadá é frágil e ameaçado, e permanece escassamente documentado. São, portanto, necessárias medidas para proteger o conhecimento sobre o património marítimo e subaquático, tanto tangível como intangível, e compreender os impactos atuais e futuros na preservação e degradação destes recursos únicos não renováveis. Os projetos da Equipa de Arqueologia Subaquática dos Parques do Canada demonstram a consideração e a importância da conservação do património natural e cultural, uma vez que a arqueologia subaquática contribui para a compreensão da relação da humanidade com o oceano, e a necessidade urgente de adaptação às alterações climáticas.

L'Organisation des Nations Unies pour l'éducation, la science et la culture a lancé la Décennie des sciences océaniques pour le développement durable (2021-2030) qui souligne la nécessité de meilleures approches collaboratives dans les sciences de la mer pour la conservation et la durabilité des océans. L'archéologie subaquatique est une science diversifiée qui a développé et participé à des projets multidisciplinaires. Compte tenu des répercussions des changements climatiques, le patrimoine submergé au Canada est fragile et menacé, et demeure à peine documenté. Des mesures sont donc nécessaires pour protéger les connaissances sur le patrimoine maritime et subaquatique, à la fois matériel et immatériel, et comprendre les impacts actuels et futurs sur la préservation et la dégradation de ces ressources non renouvelables uniques. Les projets de l'équipe d'archéologie subaquatique de Parcs Canada mettent en valeur la considération et l'importance de la conservation à la fois du patrimoine naturel et culturel, car l'archéologie subaquatique contribue à la compréhension de la relation de l'humanité avec l'océan et du besoin urgent d'adaptation aux changements climatiques.

Introduction

When the United Nations Educational, Scientific and Cultural Organization (UNESCO) launched the Decade of Ocean Science for Sustainable Development (2021–2030), which highlighted the need for collaborative approaches for ocean conservation and sustainability, it became quite clear that the fate of humankind is directly linked to the fate of the ocean. While the immediate reflex is to look at traditional marine sciences, other disciplines can also play an important role. In particular, underwater archaeology can play many key roles in supporting the global effort to improve the state of the ocean. Underwater archaeological projects offer multidisciplinary research opportunities at a time when access to sites is challenging. Perhaps, more importantly, these projects have incredible appeal to the public and can play a significant role in building engagement by enhancing the understanding of humanity's past relationship with the ocean.

Research in marine sciences should include both cultural and natural resources. Archaeological sites can shed light on past practices, for better or for worse; researchers can learn from them to understand their impact on the current condition of the ocean. Shipwrecks offer concentrated ecosystems that can be amazing research areas, but also markers for climate change. Underwater archaeology is, therefore, a vector of change and development for multidisciplinary projects. Given the impacts of climate change, submerged heritage is fragile and threatened on a short-, medium-, and long-term scale, and remains incompletely documented in Canada. Measures are, therefore, necessary to protect knowledge on maritime and underwater cultural heritage, both tangible and intangible, and understand the current and future impacts on the conservation and degradation of these unique non-renewable resources.

Parks Canada's Underwater Archaeology Team's (UAT) projects showcase how, within the dual mandate of Parks Canada to preserve natural and cultural heritage, underwater archaeology contributes to the understanding of humankind's relationship with the ocean and the need for adaptation to climate change. Using examples from the Atlantic, Pacific, and Arctic Oceans, as well as from the Great Lakes and the St. Lawrence River, this discussion will go beyond the simple need to protect archaeological remains of humanity's past and explore how underwater cultural heritage can help build a better future for the ocean by looking into the past.

International Consideration of Underwater Cultural Heritage

The *Convention on the Law of the Sea* (UNCLOS) (United Nations 1982) first started to assert the importance of archaeological and historical sites and artefacts in marine environments. Article 145 mentions the protection and conservation of the natural resources of marine environments to prevent damage to flora and fauna, and article 193 the general obligation for the States that have consented to be bound by the Convention to protect and preserve their marine environments, that could only include natural resources in general. Article 149 goes further, however, and states that "all objects of an archaeological and historical nature found in the Area shall be preserved or disposed of for the benefit of [hu]mankind as a whole" (United Nations 1982). Article 150.f authorizes activities in an area to be conducted with a view to ensure "the development of the common heritage for the benefit of [hu]mankind as a whole" (United Nations 1982). Finally, article 303.1 affirms that "States have the duty to protect objects of an archaeological and historical nature found at sea and shall cooperate for this purpose" (United Nations 1982).

A few decades later, UNESCO published the Convention on the Protection of the Underwater Cultural Heritage in 2001 (UNESCO 2001) to set out the principles and guidelines that States must follow in safeguarding of underwater cultural heritage. In 2019, for the upcoming 20th anniversary of the convention, the International Committee on Underwater Cultural Heritage (ICUCH) was mandated by the International Council on Monuments and Sites (ICOMOS) to review the guidelines and make recommendations to promote further cooperation, and better management of submerged sites by integrating marine environmental and archaeological science with the develop an open database, amongst others.

More recently, UNESCO launched the Decade of Ocean Science for Sustainable Development (2021–2030), which includes three main objectives (11, 13 and 14) that impact communities, ocean sciences, and the climate change phenomena (UNESCO 2021). The objectives do not directly mention submerged archaeological sites, but it became clear to scientific and archaeological communities around the world that underwater cultural heritage was not only a resource that needed to be considered and protected, but that this non-renewable resource could actually play a significant role in the fight against climate change. Canada is no exception to providing examples of this role.

Underwater Cultural Heritage and Climate Change

Archaeological sites, in general, are at risk from both natural and cultural impacts. In the case of terrestrial sites, the greatest risks are often related to urban development or shoreline erosion. The highest risks in the context of underwater archaeological sites are anthropic activities, such as dredging, trawling, and looting atop the backdrop of exposure to the full range of environmental conditions. Pre- and post-depositional processes, such as a fire onboard a ship or storms after a wreckage, degrade submerged or semi-submerged sites. Over time though, they eventually reach a certain equilibrium depending on the dynamics of the area.

In the context of climate change, if some of the ambient factors change, like an increase in precipitation, melting of ice cover, rise in temperature, sea-level change, or acidification, the effects or impacts can become exponential and will accelerate the destruction of the sites.

As the acceleration of these phenomena becomes urgent and forces action, the first challenge is to develop resilience, given that all sites could eventually be destroyed or severely degraded, and potentially disappear on a timescale of from one to hundreds of years from now. It also requires consideration of how to apply mitigation and adaptation measures specifically in marine environments for cultural resources (Underwood 2022).

Dual Mandate at Parks Canada and Climate Change

Parks Canada Agency is responsible for a national system of Canadian natural and cultural heritage places, including 48 national parks, and three national marine conservation areas, including 172 national historic sites, a national urban park, and a national landmark. On behalf of the people of Canada, Parks Canada's mission is "to protect and present nationally significant examples of Canada's natural and cultural heritage and foster public understanding, appreciation and enjoyment in ways that ensure their ecological and commemorative integrity for present and future generations" (Parks Canada Agency 2023). Within the Agency, climate-change initiatives include, amongst other efforts, a dedicated team in the Office of the Chief Ecosystem Scientist of Parks Canada, the Parks Canada Conservation and Restoration Program (Parks Canada 2018), an Ocean Literacy working group, a National Marine Conservation Areas (NMCA) program, and a recently updated NMCA policy aligned with the Canada National Marine Conservation Areas Act (2002) (Government of Canada 2002). Collectively, these initiatives support the development of adaptation and mitigation measures for the conservation of both natural and cultural resources. Public outreach activities, media, and social media campaigns, include being an organizer and co-host of the 5th International Marine Protected Areas Congress (IMPAC5) in Vancouver in February 2023.

Multiple examples exist from all over the country of climate change adaptation projects and workshops, including interventions in 2017 at the Dawson Historical Complex, Yukon, the York Factory National Historical Site, Manitoba, and Forillon National Park, Quebec. In 2018, other work was done at Lake Louise, Yoho, Alberta, Terra Nova National Park, Newfoundland, and Fortress of Louisbourg National Historical Site, Nova Scotia. In 2019, amongst others, projects were conducted at Bruce Peninsula National Park, Fathom Five National Marine Park, Ontario and Kejimkujik National Park and National Historical Site, Nova Scotia (Ocean Literacy Working Group 2022). Some of these example projects were solely focused on cultural heritage, others had an ecological integrity focus, and one example integrates both natural and cultural elements. Given that the cultural heritage projects concerned mainly terrestrial or coastal archaeological resources, what about underwater cultural heritage?

Underwater Archaeology at Parks Canada

Since 1964, the UAT has been the sole federal government organization with underwater cultural heritage expertise and professional scientific diving capacity. The team conducts research mainly at national historic sites, collaborates with field units that manage those sites, and selectively provides external advice for environmental impact assessments on projects that may include underwater cultural resources. Both salt and freshwater environments are of importance in Canada, with three coastlines bordering the Atlantic, the Pacific and the Arctic oceans; the five Great Lakes; multiple peninsulas; countless offshore islands, and vast riverine networks all over the country (World Atlas 2023). As the full extent of underwater cultural resources remain unknown and applied expertise is limited for such a wide geographic area, many concerns exist for the conservation of underwater heritage and the present and future impacts of climate change.

Through the lens of climate change, what insight does underwater heritage offer in Canada? How can sites be understood, and can their preservation be promoted, especially with the diverse environments all over the country? Or at least preserve the knowledge related to those sites. Furthermore, the question of how to commemorate these sites is critical, whether for their importance in history, for their scientific and archaeological values, or even for their significance for local communities, as grief, or a sense of belonging has often been developed in relation to those sites. What follows are some projects that the UAT realized or participated in, over the last few years that showcase different stages of consideration for underwater cultural heritage, environment, conservation and eventually climate change, from exclusively archaeological projects to multidisciplinary interventions.

Archaeological Projects and Impacts on Underwater Cultural Heritage

A three-year project in Ontario, at Lake Superior NMCA included sonar surveys and underwater recording on various known shipwrecks by local divers and communities, but that were not previously visited by archaeologists. Underwater archaeologists recorded various sites, such as the shipwrecks of *W.J. Emerson* (1900–1933) (Figure 1), *Neebing* (1892, sunk in 1937), and *Ontario* (1874–1899). On *Mary E. Machlachlan* (1893, sunk in 1921), for example, information was gathered on naval architecture and the integrity of the wrecks, and visual products to eventually share with the public were recorded. Observations were made on the presence of known invasive species, including the first sighting of zebra mussels (Dagneau 2023). Despite their limited numbers on the wreck, and their presence that geographically high and deep into the largest inland lake in the country, it raises questions about the arrival of more invasive species, and their future impacts on local wrecks.

FIGURE 1. Underwater archaeologists recording the wreck of the *W. J. Emerson* in Lake Superior, Ontario, Canada. (Photo by the author, Parcs Canada, 2022.)

The direct impacts on the environment could include a change in visibility, acidification of the water, and/or rising water temperature. All are factors that could accelerate the degradation of wood and metal from the wreck. There are also concerns about water level fluctuations for the Great Lakes in general (Environmental Protection Agency 2022), and the duration of the yearly ice cover. Some shallow sites can be destroyed by ice, but it can also protect deeper sites from year-long sunlight, and wind-wave actions.

Other indirect impacts to consider include health and safety, and socio economic concerns for tourism,

recreational diving, local communities, and the work of underwater archaeologists and other researchers. If the integrity of the wreck is compromised and knowledge is lost, whether on a short- or long-term basis, this could be disastrous. Therefore, many factors remain to be considered and more information to be gathered to even be able to begin to identify, quantify, and understand all the upcoming impacts of a changing environment on the wrecks in Lake Superior NMCA. These considerations will contribute to an effective management plan for underwater cultural resources.

Environmental and Archaeological Projects

Since 2020, the UAT has collaborated on both environmental and archaeological projects in the St. Lawrence River amongst others. In the Saguenay-Saint-Lawrence Marine Park, the UAT mitigated the ecological impacts of work by mounting a "double" project. This was done to optimize the equipment and resources deployed from the Research Vessel *David Thompson* (Figure 2), and also an ROV, side-scan sonar, and multibeam echosounder. The combined operation was meant to use less fuel, for example, than executing two separate projects, and also to maximize financial and human resources (Bernier and Néron 2021, 2023).

FIGURE 2. Research Vessel *David Thompson* at Ile Rouge in the Marine Park of Saguenay-St. Lawrence. (Photo by the author, Parcs Canada, 2021.)

The main objectives of this 2020–2021 project were to revisit sites that were recorded manually and with older sonars over 20 years ago, and use new technologies to gather better data. It was also possible to document some new unknown wrecks that were recently discovered in deep water by the Canadian Hydrographic Service. Observations on current integrity status of the known sites provided information on the degradation that occurred to these sites through time by layering old site plans with new sonar imagery. The environmental side of the mission included testing new instruments, such as baited cameras and current (C), temperature (T) and pressure (D) (CTD) probes (Figure 3), and establishing protocols for the biological and ecological objectives, including microorganism sampling. CTD equipment is used to detect how the conductivity and temperature of water changes relative to depth. The depth is also a marker for pressure of the seawater, and the conductivity is used to determine salinity (Sea-Bird Scientific 2023).

FIGURE 3. Underwater archaeologist retrieving a current-temperature-depth probe for environmental data. (Photo courtesy of Jonathan Moore, Parks Canada, 2021.)

It is well known how important it is to understand the environment surrounding archaeological sites, and the dynamics at play, like current, tides, microorganisms, wind, and the nature of seabed, etc., to quantify their stability, integrity, and post-depositional processes on different scales. Biodiversity is also important to be aware of and understand as living organisms co-exist with archaeological remains given that they live, interact, and have an impact on them.

Multidisciplinarity for Maritime Archaeology

Multidisciplinarity in underwater archaeological projects has become more and more important in recent years, and is clearly part of the future of this discipline and marine sciences, in general. UAT archaeological research at the wreck of HMS *Erebus* (1826, deserted in 1848) in the Arctic Ocean, Nunavut, Canada, is mainly focused on the wreck as the ship that it was, a state-of-the-art machine of its own time, an element in a political and socio-economic system to explore the North-West passage, and as a closed community that was home for a few short years for seamen and officers, alike, that all had professions, their own feelings and behaviors, and social interactions on board. The project also included biodiversity studies, given that it is an unplanned artificial reef, on top of archival research, oral history, and the inclusion of local Indigenous knowledge (Moore 2020).

New technologies have been used throughout the years as multibeam echosounders, side-scan sonar, remotely operated vehicles (ROVs), drones, photogrammetry, and 3D modeling. All these tools help to not only conduct exploration, to discover and record the shipwreck, but also to map the Arctic seabed. The data helps provide a global perspective of the site and the debris field, with higher resolution and more precise data; and it provides ways of effectively sharing these discoveries with the public and all the people who will never be able to visit those sites in person. Additionally, an ongoing study of the wind-wave spectrum of the area is progressing, and a database for the ice cover as the Arctic is melting much quicker than previously anticipated is being developed (Duggan 2023). The main goal of all those studies is to understand the archaeological remains as they were and are, and the site dynamics, to create a baseline and be able to predict future impacts on the wreck of HMS *Erebus* either from its environment and/or the different factors related to climate change.

Archaeology as an Added Value for Multidisciplinarity

In 2017, the UAT participated in an ecological/cultural project at Gulf Islands National Park Reserve, British Columbia, with the main project goal to revitalize the traditional use and management of shellfish and other marine food resources at a 'sea garden' and learn more about the benefits and impacts on local intertidal ecosystems. The project included geographical, biological, stratigraphic, and paleoenvironmental studies, and intertidal and underwater archaeological test excavations for dating evidence, in collaboration with Coast Salish

Indigenous communities. Overall, the project aim was to restore an Indigenous ecological management tool and to provide opportunities for them to reconnect with their ancestral knowledge, traditional places, and practices (Parks Canada 2022).

Concerns were also raised with the proximity of increasing maritime traffic with propellor wash and anchors impacts, the introduction of invasive species, the threats of erosion, and so on. Therefore, another objective was to understand the site and to consider upcoming impacts of climate change. The UAT provided underwater field expertise, including excavation, recording, sampling, and photogrammetry (Figure 4) (Smith and Moore 2019). Archaeology was an added project value to help understand ancient lifeways, given that intangible heritage goes hand-in-hand with tangible heritage, especially for people who lived in harmony with nature for millennia.

A Holistic Approach for Future Projects

One of the upcoming projects of the UAT will take place in Mingan, Québec, also in the St. Lawrence River. The project will try, in a holistic integrated approach, to consider all those methodologies, tools, processes, new perspectives and multidisciplinarities that were mentioned in the preceding examples. On an ecological level, the project aims to gather biological and environmental information in the area, including CTD data, local fauna and flora, the composition of the seabed, and coral sampling, for example. Archaeological objectives include revisiting a PBY 5A American military aircraft that sank in 1942, the wrecks of the SS *Clyde* (1857) and SS *North Briton* (1861), located at Wreck Island, amongst others, with newer equipment to enhance recording and tackle unanswered questions. Another objective will be to monitor the evolution of site degradation, if any, of known archaeological sites and, of course, changes in site dynamics, in an attempt to gauge site data preservation risks and loss.

Considerations for Management, Research, and Conservation

It is not only important to understand past and present underwater cultural heritage as archaeological sites, but also to understand how they exist in the present within broader complex ecosystems that are often at equilibrium. Researchers try to dissolve the dichotomy between nature and culture, because one cannot be understood without the other. Furthermore,

the sustainable development need for collaborative approaches, the heritage as a carrier of sense or meaning, and the urgency coming from climate change, scientists also need to dissolve the dichotomy between the past and the present. As of now, the future also needs to be integrated in the equation to understand how the sites are going to be impacted, destroyed, preserved, studied, commemorated, or forgotten.

FIGURE 4. Underwater archaeologist documenting Fulford sea garden, British Columbia. (Photo courtesy of Thierry Boyer, Parks Canada, 2019.)

With this information, it will then be possible to set an Underwater Strategy, with Risk Assessments for submerged archaeological sites and structures, Management Plans that will help scale the time resources have remaining, and prioritization for intervention or conservation. This can lead to the application of adaptation measures to aid preservation for future researchers to study with perhaps better technologies and new interpretations.

The Importance of Underwater Cultural Heritage for Addressing Climate Change

Why is underwater cultural heritage important in caring for the ocean and in addressing climate change? From one generation to another, an evolution has occurred in the definition of underwater cultural heritage; the sea is often viewed as a memory, even as the subconscious of a society, embodying tangible and intangible heritages that are increasingly considered as a whole. Within this milieu, underwater sites themselves are vectors of change and adaptation, and have their rightful place in the paradigm of sciences where multidisciplinarity now prevails. Underwater archaeology can build bridges between environmental and social sciences, and be used to help with adaptation and sustainable development to support a heritage fully at risk. Moreover, great public interest exists and mystery surrounds sites in the depth of ocean, lakes and rivers. Underwater cultural heritage has the ability to bring knowledge to the surface and enlighten wider issues and concerns held by many, whether it is marine conservation, climate change, or the health of the ocean, for the future generations to come.

References

BERNIER, MARC-ANDRÉ, AND AIMIE NÉRON
2021   *Archaeological Intervention at Saguenay-Saint-Lawrence Marine Park 2020 (Intervention archéologique au Parc marin du Saguenay-Saint-Laurent 2020)*. Équipe d'archéologie subaquatique, Parcs Canada, Ottawa, CA.

2023   *Archaeological Intervention at Saguenay-Saint-Lawrence Marine Park 2021 (Intervention archéologique au Parc marin du Saguenay-Saint-Laurent 2021)*. Équipe d'archéologie subaquatique, Parcs Canada, Ottawa, CA.

DAGNEAU, CHARLES
2023   *Lake Superior National Marine Conservation Area. Submerged Cultural Resource Inventory Project 2022*. Ongoing report, Underwater Archaeology Team, Parks Canada, Ottawa, CA.

DUGGAN, GRAHAM
2023   "The Arctic is warming faster than anywhere else on the planet". The Nature of Things. CBC <https://www.cbc.ca/natureofthings/features/the-arctic-is-warming-faster-than-anywhere-else-on-the-planet>. Accessed 25 January 2023.

ENVIRONMENTAL PROTECTION AGENCY
2022   Climate Change Indicators: Great Lakes Water Levels and Temperatures. United States Environmental Protection Agency. < https://www.epa.gov/climate-indicators/great-lakes>. Accessed 23 January 2023.

GOVERNMENT OF CANADA
2002   Canada National Marine Conservation Areas Act 2002. Justice Laws Website <https://laws-lois.justice.gc.ca/eng/acts/c-7.3/page-1.html>. Accessed 25 February 2023.

MOORE, JONATHAN
2020   The Wreck of HMS *Erebus*: A Fieldwork and Research Update. In *ACUA Underwater Archaeology Proceedings 2020*, Victor Mastone and Calvin Mires (editors), pp. 125–134. Advisory Council on Underwater Archaeology and Society for Historical Archaeology.

OCEAN LITERACY WORKING GROUP
2022   Climate Change at Parks Canada. Ocean Literacy Working Group Meeting. Ocean Literacy Group, Parks Canada Presentation, 4 February 2022.

PARKS CANADA
2018   *A Natural Priority - A Report on Parks Canada's Conservation and Restoration Program*. Parks Canada Agency, Ottawa, CA.

2022   Sea Garden Restoration. Gulf Islands National Park Reserve. Government of Canada < https://parks.canada.ca/pn-np/bc/gulf/nature/restauration-restoration/jardins-de-la-mer-sea-gardens>. Accessed 15 February 2023.

2023   About the Parks Canada Agency. Government of Canada < https://parks.canada.ca/agence-agency>. Accessed 16 December 2022.

SEA-BIRD SCIENTIFIC
2023   Guide to CTDs. Measure Conductivity and Temperature with an Oceanographic CTD. Sea-Bird Scientific < https://www.seabird.com/eBooks/CTDs-Explained-Sea-Bird-Scientific?>. Accessed 25 January 2023.

SMITH, NICOLE, AND JONATHAN MOORE
2019   *h*Manuscript, Nicole Smith and Underwater Archaeology Team, Parks Canada, Ottawa.

UNDERWOOD, CHRISTOPHER J.
2022   Cultural Heritage: A Driver for Transformative Change, Adaptation and Sustainability. Poster, ICOMOS-ICUCH International Committee on the Underwater Cultural Heritage.

UNESCO
2001   Convention on the Protection of the Underwater Cultural Heritage. UNESCO < https://unesdoc.unesco.org/ark:/48223/pf0000126065>. Accessed 25 March 2023.

2021   United Nations Decade of Ocean Science for Sustainable Development (2021–2030). UNESCO < https://en.unesco.org/ocean-decade>. Accessed 25 March 2023.

UNITED NATIONS
1982   *United Nations Convention on the Law of the Sea*. UNCLOS, United Nations.

WORLD ATLAS
2023   The Coastline of Canada, The Longest in The World. World Atlas. < https://www.worldatlas.com/articles/the-coastline-of-canada-the-longest-in-the-world.html>. Accessed 3 January 2023.

· · · · · · · · · · · · · ·

Aimie Néron

Agence Parcs Canada | Parks Canada Agency

1800 Walkley, Ottawa (Ontario)

Canada K1H8K3

From Shore to Shore: The Construction of Ferries in Saskatchewan, Canada

Michael K. Lewis

Prior to the construction of bridges, the most common and safest method to cross the rivers in the Canadian prairies was to be ferried across, due to the severe and dangerous currents within the rivers. These ferries were locally manufactured to no standard plan, with the knowledge that the ferries would have a limited useful life span before being discarded. This paper describes the geographic region of their service, construction methods, and includes a 3D-model based upon a preserved ferry. Thus, providing an archaeological record of these important, but easily overlooked, watercraft in the Canadian Prairies.

Antes da construção de pontes, o método mais comum e mais seguro de atravessar os rios nas pradarias canadianas era o ferry, devido às correntes fortes e perigosas que existiam nos rios. Estes ferries eram fabricados localmente sem qualquer plano padrão, sabendo-se que teriam uma vida útil limitada antes de serem descartados. O presente documento descreve a região geográfica em que se encontravam, os métodos de construção e inclui um modelo 3D baseado num ferry preservado. Deste modo, fornece um registo arqueológico destas embarcações importantes, mas facilmente ignoradas, das pradarias canadianas.

Avant la construction de ponts, la méthode la plus courante et la plus sûre pour traverser les rivières dans les Prairies canadiennes était d'être transporté d'une rive à l'autre, en raison des courants sévères et dangereux dans les rivières. Ces traversiers ont été fabriqués localement sans plan standard, sachant que les traversiers auraient une durée de vie utile limitée avant d'être abandonnés. Cet article décrit la région géographique de leur service, les méthodes de construction, et comprend un modèle 3D basé sur un traversier préservé. Ainsi, ceci nous fourni un enregistrement archéologique de ces embarcations importantes, mais facilement négligées, dans les Prairies canadiennes.

Introduction

When one thinks of the province of Saskatchewan, what commonly comes to mind is vast, flat, treeless, grassy land in the middle of the Canadian Prairies, with no bodies of water and covered in snow seven or eight months of the year. However, this is far from the whole story, except for the snow part, and maybe the flatness. Saskatchewan is one of the three prairie provinces in Western Canada, the others being Alberta (to the west of Saskatchewan) and Manitoba (to the east of Saskatchewan). It is often referred to as the province that is hard to spell, but easy to draw, as the borders form a rough rectangle (49th and 60th parallels of latitude, and 102 and 110 degrees longitude). The vast flat expanse often considered to the general appearance of what Saskatchewan looks like only applies to the southern part of the province. The northern part of the province has areas of hilly terrain, permafrost, bogs, boreal forests, and over 100,000 lakes. In addition to these bodies of water, Saskatchewan has multiple rivers that cross through the province, the most predominant of which are the North and South Saskatchewan rivers. The Saskatchewan River is a fast-flowing river, with steep banks and few, if any, fording locations. As immigration caused the population of the province to rise in the early 20th century, the province developed a ferry system to help transport people and goods through the region. While many of the ferries have been replaced with bridges or ferries of more modern construction, those original ferries played a vital role in the development of the province. This paper describes the construction of the early ferries, along with their propulsion methods and historical significance.

Saskatchewan Geography

The province of Saskatchewan, in the western Canadian Prairies (Figure 1), is a landlocked region. Saskatchewan has a total area of 651,036 square kilometers (km2) (251,366 square miles [mi.2]) of which 591,670 km2 (228,450 mi.2) is land and 59,366 km2 (22,921 mi.2) is water (University of Regina 2023). Saskatchewan contains 4 distinct ecological regions, the Taiga Shield, in the farthest North; the Boreal Shield, directly south of the Taiga Shield; the Boreal Plain, south of the Boreal Shield; and the Prairie Plain in the southern third of the province (University of Regina 2023). The Taiga Shield's terrain was developed by advancing and retreating glaciers, with either flat areas where the depressions are now lakes and surrounding areas, which are either temporarily or permanently waterlogged. Permafrost exists over a large area of the Taiga Shield, as the average annual temperature is just below freezing. This area is also a part of the Canadian Shield, with rocky terrain, and what soil available is low in nutrients and high in acid. However, vegetation still grows in these hostile conditions; the most common type of tree in this area is coniferous trees, with spruce, fir, pine and tamarack or larch being the four common species. Thus, the Taiga Shield is a mixture of wetlands, with permafrost, forests, meadows, and shrubs covering the rocky terrain.

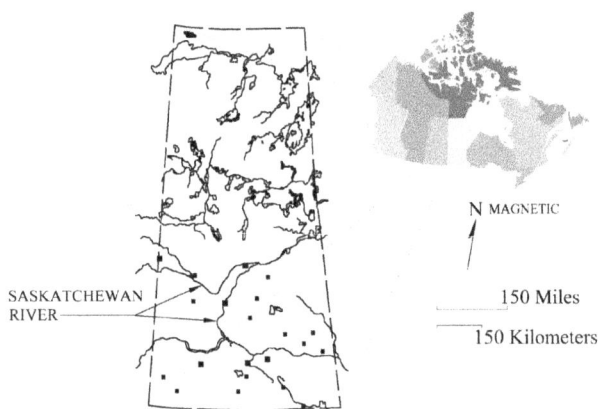

FIGURE 1. Map of Saskatchewan showing rivers and population centers (size of square dependent on population size). (Figure created by the author, 2023.)

The Boreal Shield is similar in some regards to the Taiga Shield. It is also a part of the Canadian Shield, and thus was also developed by the movements of glaciers over eons past. But the Boreal Shield has no soil on its bedrock outcroppings, just peat and sand, with other terrain being either bogs or wetlands, with permafrost still being common (University of Regina 2023). Plants also thrived in this area, with much of the region covered in dense forest made up of black spruce, with birch, balsam poplar, and jack pine growing in both wind protected and open areas (University of Regina 2023). Bogs and wetlands cover one-fifth of this area, the glacier movements that developed this area also created an extensive network of lakes and rivers. These extensive waterways, which continued into the Boreal Plains ecological area, were the roads during the fur trade, making use of the plentiful wildlife in the area, including beaver, mink, black bears to name a few (University of Regina 2023). The Boral Plains also is plentiful in wildlife. This area's terrain varies with the northern part being similar to the

Boreal Shield containing no soil, but plenty of sand and peat bogs and the southern part where the soil is able to support agriculture; however, the growing season is limited because on average only 103 days are frost free (University of Regina 2023). This area was important to the fur trade, and is the most populated region of the northern two-thirds of the province. This population lives in small communities rather than the larger urban centers found in the southern third of the province (University of Regina 2023).

The Prairie Plains begin south of the Boreal Plains, with a band of parklands, containing aspen groves, with occasional open spaces or meadows that open into typical grassland prairies to the south (University of Regina 2023). The Prairie Plains were also affected by the melting glaciers, but this movement has flattened the landscape and left fertile deposits of black soil that is still sought after for high-yield grain production (University of Regina 2023). Considering the environmental, geological, and geographical conditions of the northern two-thirds of the province, it is plain to see why the majority of the inhabitants are located in the southern one-third (University of Regina 2023). Over 80% of the population of Saskatchewan live in this region (University of Regina 2023).

As shown in Figure 1, the province is crisscrossed with rivers, with the most important being the Saskatchewan River. The headwaters of the Saskatchewan River begin in the Rocky Mountains of the Province of Alberta. As these streams and rivers flow and drain to the east, they join to become the North and South Saskatchewan rivers. In the province of Saskatchewan, these two rivers join to form the Saskatchewan River, which continues eastward to Lake Winnipeg in the province of Manitoba (Bast and Campbell 2022). With most of the settlements in the southern third of the province (University of Regina 2023) due to the location of the rivers, the province is nearly divided into east and west areas, thus requiring river crossings for the movement of both the population and goods (Light 2022).

Historical Reason for the Use of Ferries in Saskatchewan and their Continued Importance

Bridges require significant funding for both construction and maintenance; therefore, finding ideal locations require time to study population movement. However, Saskatchewan grew from less than 100,000 people in 1900, to over 900,000 by 1930 (University of Regina 2023), which meant the population grew faster than it was practical to build bridges. In 1905, the year Saskatchewan became a province, 18 Ferries were in operation. At the peak of the ferry system in 1929, 48 ferries were in operation. By 1960, these numbers had decreased to only 28 ferries in operation. The cause of this decrease in ferries was solely due to the construction of bridges in areas that showed high traffic volumes, which would justify the expense of construction of the bridge and its continued maintenance (Light 2022). The Government of Saskatchewan still operates 12 ferries and one barge in locations that do not have sufficient volume of traffic to warrant the construction of a bridge (Government of Saskatchewan 2023). Three of these ferries are on the North Saskatchewan River, nine of these ferries are on the South Saskatchewan River and the barge is on Wollaston Lake, which is approximately 850 kilometers (km) (528 miles [mi.]) north-east of Saskatoon (the largest population center in the province). These remaining ferries are either in extremely remote locations, such as the Wollaston Lake barge, or have bridges within a reasonable distance for winter crossings, such as the Clarkboro and Hague ferries. The current ferries typically operate from April to November, but operations depend on weather and whether the river has frozen over. During the operating season, fluctuating water levels in the rivers could cause ferry operation and maximum weights allowed to change without notice (Government of Saskatchewan 2023). Many Saskatchewan children have fond memories of their parents taking them for a drive in the summer to experience one of these ferry crossings.

Ferry Construction and Operation

This project began as a family discussion about the Frenchman Butte ferry, which was in operation from 1913 to 1968. While no construction information specific to the Frenchman Butte ferry exists, all the ferries at that time were constructed in a similar manner. A construction crew of seven or eight men (including the foreman), of whom only the foreman had any knowledge of ferry construction, travelled to all the ferry sites in the province constructing new, repairing, or rebuilding, as needed, shown in Figure 2a (Light 2022). Minor repairs and routine maintenance, such as plugging cracks and spaces between boards with oakum and pitch prior to launching for the spring, were done by the ferrymen (operators). The ferrymen also determined when the last crossings would be for the winter and pulled the ferry from the water to protect it from the ice. The case study

for this project was of a replica wooden hull ferry on display at the Western Development Museum (WDM) in Moose Jaw, Saskatchewan.

FIGURE 2. Construction of the wooden ferry hull, a) rotating the hull during construction (family photo in the possession of the author); b) framing the scow's hull. (Figure 2b from Light 2022:63, reproduced with permission.)

Until 1964, all ferry hulls were constructed of wood. Afterward, their construction was changed to metal hulls to save costs on the lifespan of the vessel (Light 2022). Prior to this switch in construction material, the life span of wooden-hull ferries typically was eight to ten years, depending upon the amount of vehicle and foot traffic carried; available construction materials, such as unseasoned lumber; and environmental factors, such as floating debris, floods, and early or late freezes (Light 2022). In 1916, a typical ferry was 42 ft. by 16 ft., with 8 ft. aprons (12.80 meters [m] by 4.88 m, with 2.44 m aprons). Larger scows were constructed in later years to handle heavier grain traffic. These were 52 ft. by 18 ft., with 12 ft. aprons (15.85 m by 5.49 m, with 3.66 m aprons) and could carry four heavy wagon loads per trip, or six typical passenger vehicles of the era (Light 2022).

The wooden-hull ferries were of a simplified construction based upon a scow hull, consisting of flat sides, tapered ends, and flat bottomed. The scow was developed as a flat-bottomed barge, which allowed them to navigate shallow coastal waters and inland waterways

(Chapelle 1981). The flat-bottom design required the use of an internal keel, to which the frames were attached, followed by internal bracing (Chapelle 1994). This simple, but efficient and effective, construction allowed the vessels to be constructed fairly rapidly, with a relatively inexperienced team. The box shape further eased the construction because the lumber was of common dimensions and did not require much modification before being attached. The overhang provided by the tapered ends allow the vessel to be beached in the shallow water while either loading or unloading from dryland. Figure 2b shows the simplified construction used on the hull. A wooden deck was constructed on top of the scow's hull, with a tiller on a platform projecting outward on the upriver side. A matching platform on the downriver side was used to carry the rowboat used in case of emergencies (Light 2022). A wooden railing was constructed on the up and down river sides of the ferry. The shore sides of the ferry were open, with an apron (ramp) secured to each end. The addition of aprons on the river ferries allowed the wheeled traffic to load and unload at the ferry crossing. These aprons were balanced so that an operator could raise the apron by stepping on one of the apron arms to lift it into travel position. The aprons were secured when the ferry was in motion (the underway or lifted position) thru metal bars inserted into the posts supporting the rails. The apron was lowered into a loading/unloading position through gravity.

The method of propulsion for the ferries was uniquely suited to river crossings. The ferry was propelled across the river by directing the river's current against the current boards (Figure 3a), thereby propelling the ferry forward. The ferry was held steady using guidelines attached to the tiller (Figure 3b). The guidelines were secured to the tops of two wooden towers, constructed on either side of the riverbank in line with each other. This guideline was used to control the ferry, and to provide the resistance against strong river current that resulted in the propulsion of the ferry across the river. This entire process is shown in Figure 3c (Light 2022). This propulsion method was feasible due to the strong currents in the Saskatchewan River system. While the flow rates in the river system can vary significantly through the year depending on the weather and the runoff from the Rocky Mountains, parts of the river system can see flow rates as high as 500 cubic meters per second (m3/s). The Saskatchewan River has a mean flow rate of 280 m3/s (Saskatchewan Water Security Agency 2023).

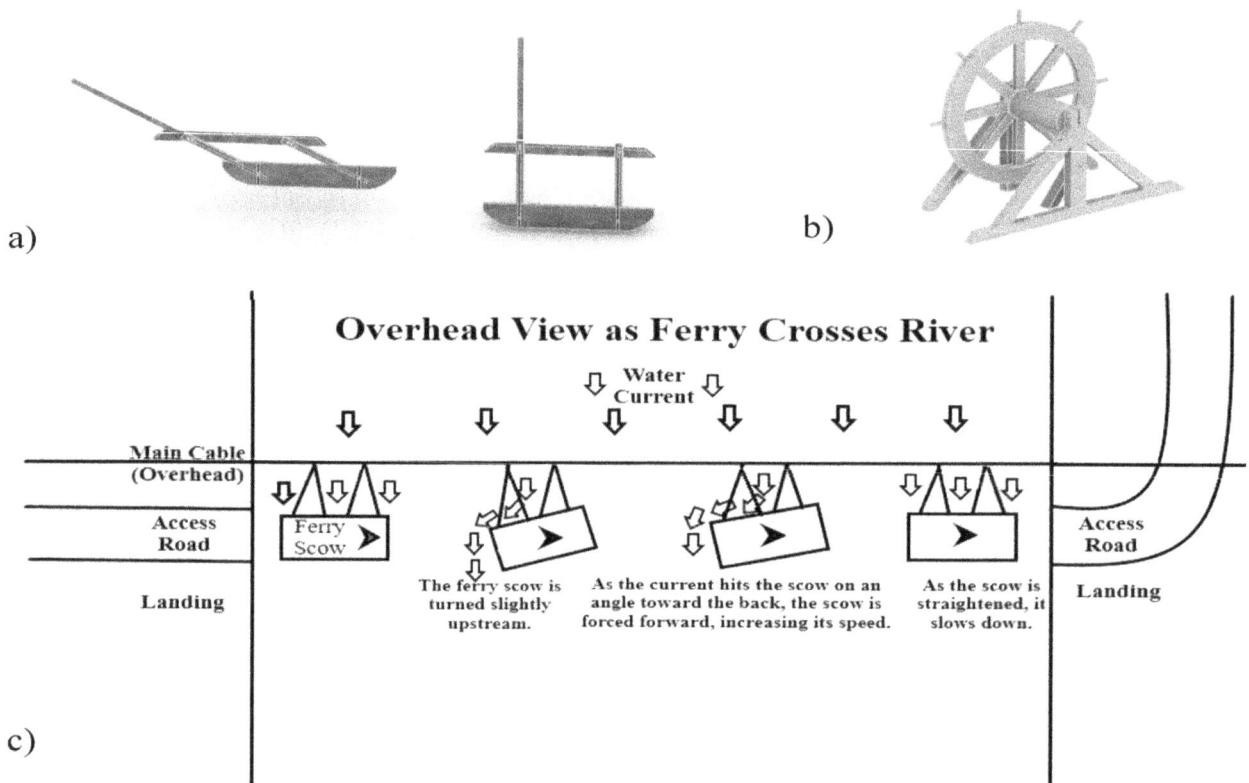

FIGURE 3. Propulsion System: a) Current boards ; b) tiller; c) ferry propulsion method. (Figures 3a and 3b created by the author, 2022; Figure 3c from Light 2022:49, reproduced with permission.)

3D-Model of the Ferry

Measurements of the replica wooden ferry located at the Western Development Museum in Moose Jaw, Saskatchewan, were taken 12 March 2022, by the author, Helen Streeton, Ruth Streeton, and Carol Light, with the assistance of the staff at the museum. The measurements were developed into both plans (Figure 4a and 4b) and a 3D-model of the ferry using Rhino CAD software. Figure 4c shows the upstream side of the ferry, including the tiller. Figure 4d shows the opposite, downstream side, including the current boards.

Further Research and Conclusion

In the future, the author is continuing research on Saskatchewan River ferries by developing a museum exhibit that will include a physical scale model, as well as the digital 3D-model of the ferry constructed for this project. This exhibit will be displayed at local museums throughout the province to help educate the public about the importance of the ferry system to the settlement and development of the province of Saskatchewan. Some museums that will be contacted about exhibition include the Western Development Museum (Moose Jaw), the Diefenbaker Center (Saskatoon), and the Frenchman Butte Museum, which is located near the site of the original Frenchman Butte ferry.

Acknowledgements

I would like to thank the staff at the Western Development Museum (WDM) Moose Jaw, Saskatchewan, for allowing us access to their replica wooden ferry to take measurements. For 39 years, the Frenchman Butte ferry was operated by Roy Sidwell, which makes the operation of the Frenchman Butte ferry part of my family story. I wish to especially thank Carol Light for her research and her recently published book *Crossing the River: A Cable Ferry History*. Without her detailed book, this project would not have been possible. I would like to thank Ruth Streeton for her diligent note and measurement taking and starting the discussion that inspired this project. Finally, I also wish to thank my wife, Helen Streeton, for her support, encouragement, and editing of this paper.

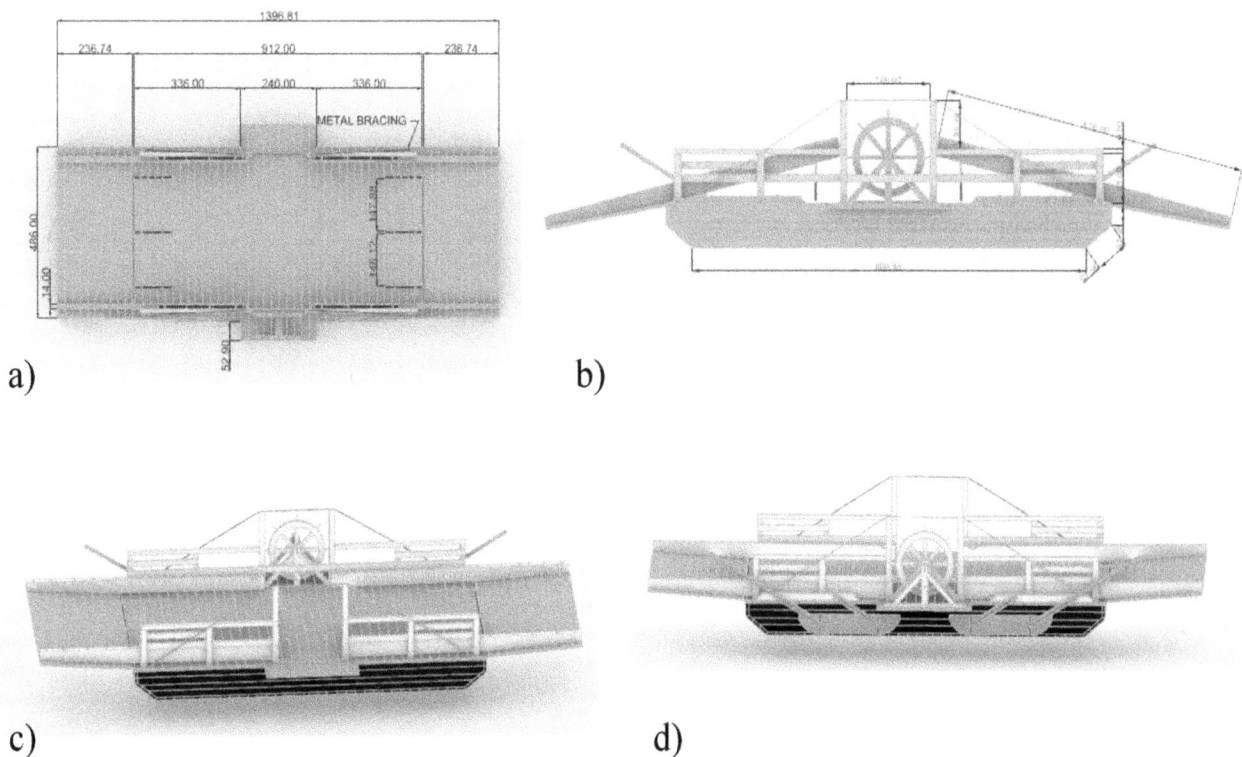

FIGURE 4. a) deck plan of ferry; b) upstream side view of ferry; c) downstream side of the ferry; d) upstream side of the ferry. (Figure created by the author, 2022.)

References

Bast, Marcy., and Ron Campbell
2022   Saskatchewan River Sturgeon Management Board
        Website. SaskPower, Regina, SK. <http://www.
        saskriversturgeon.ca/contact_us.html>. Accessed 01
        March 2023.

Chapelle, Howard Irving
1981   American Small Sailing Craft, Their Design,
        Development, and Construction. W.W Norton &
        Company, New York, NY.

1994   Boatbuilding: A Complete Handbook of Wooden
        Boat Construction. W.W Norton & Company,
        New York, NY.

Government of Saskatchewan
2023   Government of Saskatchewan Website.
        Government of Saskatchewan, Regina, SK.
        <https://www.saskatchewan.ca/residents/
        transportation/ferry-crossings> Accessed 01 March
        2023.

Light, Carol
2022   Crossing the River: A Cable History. Priority
        Printing, Edmonton, Alberta, CA.

Saskatchewan Water Security Agency
2023   Saskatchewan Water Security Agency, Government
        of Saskatchewan, Regina, SK. <https://www.wsask.
        ca/hydrographs/saskatchewan-river-watershed/>
        Accessed 01 March 2023.

University of Regina
2023   University of Regina. "Encyclopedia of
        Saskatchewan" Canadian Plains Research Center,
        Regina, SK. <https://esask.uregina.ca/about_
        encyclopedia.jsp> Accessed 28 April 2023.

· · · · · · · · · · · · · · · ·

Michael K. Lewis MSC, RPA

Director/Conservator

Conservation Of Archaeological

Materials Laboratory Saskatoon

Saskatchewan, Canada

The *Lake Austin* and the Bob Hall Pier Wrecks: A Study of Beached Shipwrecks Along Mustang and North Padre Islands, Texas

Hope Bridgeman, Hunter W. Whitehead

Historic maritime activity along the Texas coast began with Spanish exploration in the 1500s and continued over the next 500 years. Exploration, maritime shipping, fishing, shipbuilding, and tourism increased with coastal and port development, which resulted in hundreds of documented and, oftentimes, undiscovered shipwrecks within state waters and along Texas beaches. Reported beached shipwrecks, such as Lake Austin *(THC Shipwreck No. 992) and the Bob Hall Pier Wreck (THC Shipwreck No. 2459), and local wreck lore of Mustang Island and North Padre Island are presented here. Also discussed are various coastal processes affecting the preservation potential of these coastal shipwrecks.*

A atividade marítima histórica ao longo da costa do Texas começou com a exploração espanhola em 1500 e continuou durante os 500 anos seguintes. A exploração, a navegação marítima, a pesca, a construção naval e o turismo aumentaram com o desenvolvimento costeiro e portuário, o que resultou em centenas de naufrágios documentados e, muitas vezes, não descobertos em águas estatais e ao longo das praias do Texas. São aqui apresentados os naufrágios encalhados Lake Austin *(THC Shipwreck No. 992) e Bob Hall Pier Wreck (THC Shipwreck No. 2459), e a história local dos naufrágios de Mustang Island e North Padre Island. Também são discutidos os vários processos costeiros que afetam o potencial de preservação destes naufrágios costeiros.*

L'activité maritime historique le long de la côte du Texas a commencé avec l'exploration espagnole dans les années 1500 et s'est poursuivie au cours des 500 années suivantes. L'exploration, le transport maritime, la pêche, la construction navale et le tourisme ont augmenté avec le développement côtier et portuaire, ce qui a entraîné des centaines d'épaves documentées et, souvent, non découvertes dans les eaux de l'État et le long des plages du Texas. Les épaves échouées signalées, telles que le Lake Austin *(naufrage n ° 992 de la THC) et l'épave du quai Bob Hall (naufrage n ° 2459 de la THC), et les traditions locales d'épaves de l'île Mustang et de l'île North Padre sont présentées dans cet article. Il est également question de divers processus côtiers affectant le potentiel de préservation de ces épaves côtières.*

Introduction

Beached shipwrecks and treasure folklore are integral to the cultural landscape and identity of the Coastal Bend region of Texas (Figure 1). Large storm events of the 1960s uncovered several shipwrecks along Padre Island. The following discusses the Gulf scow schooner, *Lake Austin*, which eroded out of the beach on Mustang Island in 1966; and the Bob Hall Pier Wreck, discovered by treasure hunters on North Padre Island after Hurricane Beulah in 1967. *George Lincoln*, on North Padre Island (Dillon 2004), and the wreck known as Boca Chica Shipwreck No. 1 on South Padre Island (Borgens [2024]), were also discovered post-Hurricane Beulah. The state of Texas did not yet have laws protecting antiquities on public lands until the 1969 enactment of the Antiquities Code of Texas (ACT). Of note, the ACT stemmed directly from a lawsuit involving the salvage of the 1554 Spanish Plate Fleet (Arnold 1983).

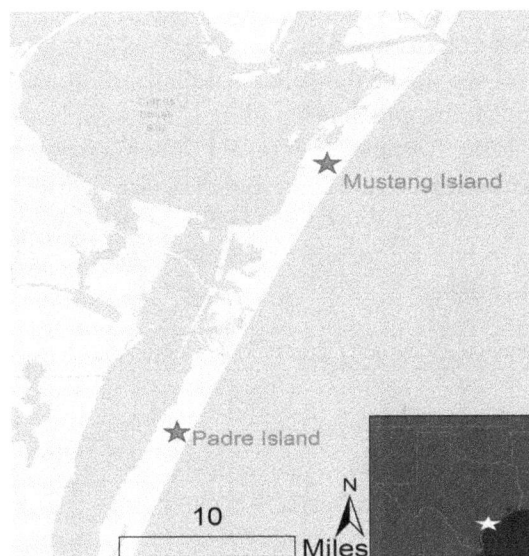

FIGURE 1. Map of the Corpus Christi Region illustrating North Padre Island and Mustang Island. (Figure by the author, 2023.)

Clearly, the 1960s were a notable point in time for the discipline of underwater archaeology in Texas.

Padre Island, and the surrounding coastal area, were popular tourist destinations by the 1920s (Scurlock 1974). Newspapers, tourism guides, souvenir maps, and beachcombing pamphlets promoted treasure hunting in the region to draw vacationers to newly constructed hotels and restaurants. Titles such as *The Secrets of Padre Island* (Smylie 1964a) and *This is Padre Island, Land of Fantasy, Fun, Treasure and Adventure* (Smylie 1964b) encouraged increased tourism and activity on the beaches. The surge in foot traffic led to the discoveries of several shipwrecks and, often, the destruction and looting of their hulls and any extant cargo. A 1938 Master's thesis, *The History of Padre Island*, notes several shipwrecks along the islands, including the tugboat *Merrimac*, cargo ship SS *Nicaragua*, and two unidentified wooden hulls (Reese 1938). Cultural processes have impacted the shipwrecks discussed here and, perhaps, most coastal wrecks to some degree.

Like many other beached shipwrecks along the Texas coast, *Lake Austin* and the Bob Hall Pier Wreck have minimal documentation. Dynamic geomorphic processes often rebury wrecks along the barrier islands shortly after they erode out of the beach, preventing archaeological investigation. Wreck exposure inextricably allows degradative physical, chemical, and biological processes to negatively impact site integrity (Keith 2016). As of 2022, the Texas Historical Commission's (THC) Marine Archaeology Program (MAP) Shipwreck Database contains 1,902 submerged archaeological sites, including shipwrecks, submerged airplanes, and portions of historic settlements (Borgens 2022). A majority of reported shipwrecks lie submerged along the Gulf of Mexico coastline including 12 recorded beached shipwrecks (Borgens [2024]). The following discusses the maritime historical context of the Corpus Christi region, the *Lake Austin* and Bob Hall Pier wreck sites, the efforts to relocate these sites, and the coastal processes affecting them.

Brief Historical Contex

The first documented depiction of the Texas coastline was drawn by Alonso Álvarez de Piñeda, who commanded a Spanish expedition of the northern Gulf of Mexico in 1519 (Chipman 1995). Piñeda named the barrier island, known today as Padre Island, Isla Blanca, which is the longest barrier island in the world. The island runs 113 miles (mi.) (182 kilometers [km]) from Port Isabel at the southern end to Corpus Christi in the north (Weymer et al. 2015). Since Piñeda's expedition, increased maritime activity has resulted in numerous shipwrecks along the coast of Texas. In 1554, three of four vessels transporting 300 passengers and crew, and cargo from Mexico to Spain ran aground on Padre Island during a storm. Months after the wreck occurred, salvage crews sent from Veracruz recovered some of the ships' cargo, including tools, guns, precious metals, and coins. This event is the first documented occurrence of Europeans on the island (Arnold and Weddle 1978).

Minimal Spanish activity occurred along Padre Island until Spanish authorities received reports that the French had established a settlement within their territory. In 1685, French explorer, René-Robert Cavelier, Sieur de La Salle, attempted to establish a colony at the mouth of the Mississippi River, arriving instead at the entrance of Matagorda Bay, where he established Fort St. Louis. La Salle was later killed by his own men in 1687 while on an expedition, and Fort St. Louis was abandoned. In 1686 and 1687, Alonso de León led two expeditions to locate and remove the French, and to reestablish the Spanish ownership of the area (Weddle 1985). De León's attempt to find the settlement was unsuccessful, as was a separate expedition in 1686–1687 commanded by Captain Martín de Rivas. Tensions between the French and Spanish over the northern Gulf of Mexico territory continued into the 18th century.

During the Mexican American War, 1846–1848, General Zachary Taylor and his troops arrived in the area in preparation for military engagements near the Rio Grande River. General Taylor encamped his troops along the southern shore of Corpus Christi Bay. The encampment at Corpus Christi held from 1,500 to about 4,000 troops prior to their deployment to Palo Alto and Resaca de la Palma (Payne 1970). Most of the battles during this war were fought on land; however, ships were used to transport troops and supplies to the coastal areas of Texas and Mexico.

During the early years of the Civil War, the entire coast of Texas was blockaded by the Union's West Gulf Blockading Squadron (Browning 2015). The transport of food, supplies, artillery, and Confederate troops in the Corpus Christi region and along the Gulf of Mexico were stifled by blockades, in a Union attempt to hobble the southern economy (Sheire 1971). In 1862, the blockader *Afton* arrived off Aransas Pass; Union forces landed on Mustang Island and burned several properties. In August, the Union attempted to take Corpus Christi; several Confederate ships were burned to prevent capture by the invading forces. In December, a military engagement occurred between the Confederate schooner, *Queen of the Bay*, and the Union screw-steamer *Sachem* and the bark *Arthur*. The Union ships were scouting along Padre Island when they discovered the Confederates attempting to check the depths of Corpus Christi Pass. The event was a minor confrontation; however, the Confederates succeeded in commandeering two Union open boats launched from *Arthur* (Browning 2015).

The depths of the Corpus Christi Pass, bays, and surrounding waters typically required vessels with a shallow draft. During the late 1800s, cattle and cotton were regularly shipped out of the Corpus Christi region on sidewheel steamers built specifically for that purpose. Charles Morgan owned a multitude of these vessels that operated along the Texas coast, including *Mary*, which sank in Aransas Pass during a storm in 1876 (Pearson and Simmons 1995). Throughout the 19th century, the various wars and lack of dependable deepwater channels within the bay impeded maritime commerce in the area until the development of the Corpus Christi Pass and the opening of the Corpus Christi harbor in 1926 (Alperin 1977). Today, vessel traffic through the Corpus Christi Shipping Channel includes large tankers, barges, and other shipping vessels, and, thus, maritime activity remains an important aspect of social and economic life in the region.

The *Lake Austin* Shipwreck

Lake Austin (THC Shipwreck No. 992), a Gulf scow schooner, was built in 1881 in Matagorda, Texas, as a shallow-draft cargo vessel that carried goods between ports in Texas and Louisiana. The vessel survived the hurricane at Indianola in 1866, served as a rescue vessel during the Great Galveston hurricane of 1900, and was later beached on Mustang Island near Port Aransas on 20 November 1903, when an intense storm caused damage to the vessel (Borgens and Bridgeman 2022). It is unknown where exactly the ship made landfall on the island. *Lake Austin*'s cargo, mainly timber, was removed from the wreck and salvaged primarily to build houses as the island had little access to timber (*Galveston Daily News* 1903; *Corpus Christi Caller-Times* 1966a). Little mention of the wreck occurred in the following decades until wave action partially uncovered *Lake Austin* throughout the early 1960s.

Believing the wreck to be a Spanish ship, Nueces County Park staff began excavating the vessel in April 1966 and soon discovered the registration number identifying the wreck (*Corpus Christi Caller-Times* 1966b; Johnson 1966). County Commissioner Melvin Littleton, the Nueces County Historical Society, and a local marine surveyor thoroughly excavated the ship by May. Plans to exhibit the wreck as a tourist attraction eventually failed due to lack of funding (*Corpus Christi Caller-Times* 1966a; 1966b; 1966c; 1966d), and two to three months later the wreck was deemed a hazard to beachgoers. As a result, what remained of the ship was broken up and burned (Johnson 1966).

Recent historical and archival research of the wreck for the THC blog (Borgens and Bridgman 2022) revealed an approximate location of the *Lake Austin* wreck during the 1966 excavation. The wreck was located between the Horace Caldwell pier and where Avenue G meets the beach road. The THC MAP files were updated, and a site form was submitted for Lake Austin; the archeological site, though no longer present, was issued state trinomial, 41NU393. Additionally, an ongoing effort to create a reproduction of *Lake Austin* is underway through the efforts of the Port Aransas Museum, in concert with Farley Boat Works. The construction of the replica ship, *Lydia Ann*, began in 2001, and is now in the later stages of production (Figure 2).

FIGURE 2. Current status of the replica ship, Lydia Ann, at Farley Boat Works. (Photo by the author, 2022.)

The Bob Hall Pier Wreck

Treasure hunters discovered the Bob Hall Pier Wreck (site 41KL108, THC Shipwreck No. 2459) on North Padre Island in fall 1967 after the effects of Hurricane Beulah uncovered a 20 foot (ft.) (6 meter [m]) portion of the wreck. The wreck was nicknamed for Bob Hall Pier by the THC as this was the landmark used to describe the wreck location in local papers—it is 3.0 mi. (4.8 km) north of where the wreck was discovered. Treasure hunters, Edmond Page, Gene French, and Richard Clement used metal detectors to further uncover portions of the wreck. Clement is quoted in a local newspaper: "There's something down there. There has to be to make all those metal detectors ring and go haywire" (*Corpus Christi Caller-Times* 1967a). The group reportedly collected three human bones, five pounds (lbs.) (2.27 kilograms [kg]) of oxidized silver, beeswax, a Roman coin, and river rock from the shipwreck. The current location and condition of these artifacts is unknown. Wooden pegs and spikes were also noted during the dismantling of the wreck during salvage efforts; a photograph of Clement holding a wooden peg in front of the wreck is shown in Figure 3.

FIGURE 3. Richard Clement holds up a six-inch wooden peg uncovered from the ribs of an old shipwreck he discovered on Padre Island shortly after Hurricane Beulah. (*Corpus Christi Caller-Times* 11 October 1967, reprinted with permission.)

Documentary research by State Marine Archaeologist, Amy Borgens, revealed newspaper accounts that suggested the shipwreck may predate the 19th century, was approximately 75 ft. (22 m) in length, contained cannon ports, consisted of two decks, and exhibited double planking. The hull construction reportedly included lead sheeting [sic] fastened with bronze nails (*Corpus Christi Caller-Times* 1967a, 1967b, 1967c, 1967d, and 1967e). This description, however, is based on local informants and is unreliable.

Whether these men had any legal right to claim the wreck, which was on state lands, caused some debate among local attorneys and the Texas attorney general's office (*Corpus Christi Caller-Times* 1967e). Efforts to salvage the ship were postponed at the request of the

Texas attorney general's office, while they determined applicable ownership laws. Ownership of shipwrecks on state lands has since been resolved in favor of Texas. Clement believed that the wreck was from the 1500s or 1600s, and claimed that a chemical analyst told him that the ship's timbers were of cedar and some of the square nails were bronze (*Corpus Christi Caller-Times* 1967a). Paul Harris, Nueces County parks manager at the time, believed the wreck may be the same vessel that was uncovered by hurricane Carla in 1961. Albert Heine, director of the Corpus Christi Museum of Science and History, viewed some of the materials recovered and commented that the wreck was likely built after 1800, probably in the latter part of the century. State archaeologist, Curtis Tunnell, is reported to have visited the wreck site to determine the ship's age, but the sand had reburied the wreck by the time he arrived. With so little information, it is difficult to determine the age of the wreck, though the use of cupreous fasteners suggests a construction date from the late 18th to 19th centuries. The treasure hunter-led excavation halted when the wreck was reburied by 16 October 1967, and the site has not been uncovered since.

In February 2021, Coastal Environments, Inc. (CEI), conducted a reconnaissance archaeological survey in search of the wreck (Whitehead 2021). This work was conducted relative to initial due-diligence cultural resources investigations of a 3,800 acre (ac.) (1,537 hectare [ha]) parcel of Nueces County Coastal Parks (NCCP) property. The pedestrian metal-detector survey did not detect any significant fields of ferrous material indicative of a shipwreck. The only detections were small and isolated. They are suggestive of marine debris that has washed in from the Gulf of Mexico or trash left behind by beachgoers. It was recommended that a combination of pedestrian and aerial drone magnetometer surveys be implemented to discover the Bob Hall Pier Wreck and any other unidentified shipwrecks within NCCP property.

Coastal Processes

Barrier islands are fundamentally dynamic; the modern composition of the Texas coast reflects competition between fluvial-deltaic sedimentation and reworking by marine geomorphic processes that produce a dominantly westerly flow in Gulf of Mexico currents. The barrier islands are also easily altered through small-scale, sea-level fluctuations; daily tidal changes; and large-scale events, such as major tropical storms and hurricanes.

Texas beaches that were previously stable are beginning to erode on a long-term basis. As Robert Morton (1994) posits, most Texas barrier islands are entering a new phase in their evolutionary history due to recent increases in the rate of sea-level rise and decreases in sediment supply. Central and North Padre Island is an exception to this trend, which he categorizes as aggradational, indicating general accumulating sand through geomorphic processes. The sand supplied by converging littoral currents along North Padre Island keeps the landform stable. This process may indicate that the Bob Hall Pier Wreck, what remains of *Lake Austin*, and potentially other shipwrecks, could be under considerable sediment and, thus, difficult to detect.

Several studies of barrier island shipwrecks have focused on the dynamic site formation processes of environmental energy and exposure changes (Horrell 2005; Russell 2005; Meide et al. 2001). Ford et al. (2016) note that barrier islands may accumulate more shipwrecks than stable shorelines because they are constantly changing due to littoral drift and aeolian processes, like erosion, transportation, and deposition. Similar processes have been observed in Franklin County, Florida, on Dog Island, where the hull remains of *Vale*, which was lost in 1899, are now completely buried due to island transgression. Alternatively, a few hundred meters away, the remains of the fishing vessel *Pricilla* have been increasingly exposed as the island moves away from the wreck (Meide et al. 2001; Ford et al. 2016).

While some beached shipwrecks may quickly degrade, scatter, or become prey to opportunistic salvage attempts, others are quickly buried by sediment, allowing for considerable preservation potential. In a case study focused on three shipwreck scatters in the Channel Islands National Park in California, Russell (2005) identified the key terminologies that archaeologists use to describe beached shipwreck sites. Delgado and Murphy (1984) used a variance of the term "buoyant hull" to characterize different types of wreckage deposited along beaches. Agranat (1994) used the term "transient shipwreck fragments" to describe segments of sunken wooden-hulled vessels that have washed ashore. One example of how quickly beached shipwreck materials can be dispersed due to environmental processes is described in Horlings' (2011) observations of a fishing vessel washed ashore in Elmina, Ghana, in 2007. For four months, the wreck was observed to break apart and move positions due to natural processes; by the end of the 2007 field season, only the bow and keel remained on shore.

Given that so little archaeological data was recorded during the wreck's exposure in 1967, it is difficult to discern how the Bob Hall Pier Wreck was deposited on North Padre Island. It is possible that the ship's hull may have been previously submerged and washed ashore during a post-depositional storm surge. From 1800 to 1900, a total of 43 recorded hurricanes made landfall in the Texas coastal area (Roth 2010), any of which could have caused the initial wrecking event or post-wrecking movement. Without first locating the Bob Hall Pier Wreck, any discussions on site formation processes are merely conjecture. In the case of *Lake Austin*, the wrecking event is well documented, and the process that would eventually see the demise of the wreck is certainly cultural.

Conclusions

Finding *Lake Austin* is unlikely due to the degree to which the vessel was burned and destroyed, yet remnants of the site may still be detectable via magnetic signature from remaining metal bolts, nails, spikes, fittings, and any other metal that was left on the ship. Without a better understanding of site location and recent geomorphic processes, it may be difficult to rediscover the Bob Hall Pier Wreck. The wreck could be underwater or potentially buried under such a considerable amount of sediment that a detectable magnetic signature may be difficult. Public outreach to determine if members of the community remember the whereabouts of the wreck would help narrow down its location, but until a systematic magnetometer survey can be undertaken, the exact location will remain a mystery.

Acknowledgements

We would like to thank the staff and volunteers at the Port Aransas Museum, and especially Director Cliff Strain for time assisting with in-person research and giving insights into the history of Mustang and Padre Islands. We would also like to thank the staff at the Corpus Christi La Retama library working in the Local History Archives for their time sifting through records for us. Lastly, we would like to thank Texas State Marine Archeologist Amy Borgens for her comments and suggestions throughout our research.

References

AGRANAT, B. J.
1994   Final Proposal: Beach Shipwreck Tracking and
        Recording, Fire Island National Seashore and
        Gateway National Recreation Area – Breezy Point
        and Sandy Hook Units: Pilot Project. Ms. on file,
        Fire Island National Seashore, Ocean Beach, CA.

ALPERIN, L.
1977   *Custodians of the Coast.* Galveston District, United
        States Army Corps of Engineers, Galveston, TX.

ARNOLD, J. BARTO
1983   The Platoro Lawsuit: Episode III Including
        the Appeals Court Decision, New Legislation,
        Permission to Sue the State and the Supreme
        Court Appeal. In I*n Search of Our Maritime
        Past: Proceedings of the Fifteenth Conference on
        Underwater Archaeology*, editors, Jonathan W.
        Bream, Rita Folse-Elliott, Claude V. Jackson, III,
        and Gordon P. Watts, Jr., pp. 82–86.

ARNOLD, J. BARTO, AND ROBERT S. WEDDLE
1978   *The Nautical Archaeology of Padre Island: The
        Spanish Shipwrecks of 1554.* Academic Press, New
        York, NY.

BORGENS, AMY A.
2022   At the Precipice of Change: 50 Years of Underwater
        Resource Management at the Texas Historical
        Commission. In *ACUA Underwater Archaeology
        Proceedings 2022*, edited by Sarah E. Holland and
        Paul F. Johnston. Advisory Council on Underwater
        Archaeology and Society of Historical Archaeology,
        45–53.

[2024] Don't Come and Take It: Chronicling
        Environmental and Anthropogenic Impacts on
        Historic Shipwrecks in the Texas Coastal Zone. In
        *The Intertidal Shipwreck: Management of a Historic
        Resource in an Unmanageable Environment*, edited
        by Jennifer E. Jones, Calvin Mires, and Daniel
        Zwick. University Florida Press, Gainesville, FL.

BORGENS, AMY A., AND HOPE BRIDGEMAN
2022   Stormed-Tossed Ships: *Lake Austin* and Her "Sister"
        Ship *Lydia Ann*. Texas Historical Commission
        <https://www.thc.texas.gov/blog/storm-tossed-
        ships-lake-austin-and-her-sister-ship-lydia-ann>.
        Accessed 5 January 2023.

BROWNING, ROBERT M.
2015   *Lincoln's Trident: The West Gulf Blockading Squadron
        During the Civil War.* United States: University of
        Alabama Press, Tuscaloosa, AL.

CHIPMAN, DONALD E.
1995   Alonso Alvarez De Pineda and the Río De Las
        Palmas: Scholars and the Mislocation of a River.
        The Southwestern Historical Quarterly 98(3):369–
        385.

CORPUS CHRISTI CALLER-TIMES
1966a   Wrecked Schooner's Cargo Built Port Aransas
        Home. *Corpus Christi Caller-Times* 19 June 1966,
        Corpus Christi, TX.

1966b   Old Schooner '*Lake Austin*' Due Berthing in
        County Park. *Corpus Christi Caller-Times* 21 May
        1966, Corpus Christi, TX.

1966c   Workers Salvage Old Wreck. *Corpus Christi Caller-
        Times* 29 April, 56 (255):1. Corpus Christi, TX.

1966d   Vessel Floated. *Corpus Christi Caller-Times* 2 May
        1966, Corpus Christi, TX.

1967a   Hope for Treasure Keeps 3 Working. *Corpus Christi
        Caller-Times* 11 October 1967, Corpus Christi,
        TX.

1967b   Treasure Hunters: Lawyer Says 3 Have Right
        to Any Claims. *Corpus Christi Caller Times* 12
        October 1967, Corpus Christi, TX.

1967c   State Interested: Official Inspects Old Ship. *Corpus
        Christi Caller-Times* 15 October 1967, Corpus
        Christi, TX.

1967d   Attempt to Salvage Ship Wreckage on Island
        Delayed. *Corpus Christi Caller-Times* 16 October
        1967, Corpus Christi, TX.

1967e   Man Plans to Date Old Ship. Corpus Christi
        Caller-Times 17 October 1967, Corpus Christi,
        TX.

DELGADO, JAMES P., AND LARRY E. MURPHY
1984   Environmentally Exposed Shipwreck Remains:
        Implications for a Natural Site Formation Process.
        Paper presented at the 15th Annual Conference on
        Underwater Archaeology, Williamsburg, VA.

DILLON, DOROTHY HOWE
2004   Shipwreck of the *George Lincoln* Between Mustang
        Island and Padre Island. Self-published. On file
        with the Corpus Christi La Retama Library, Corpus
        Christi, TX.

FORD, BEN, CARRIE SOWDEN, KATHERINE FARNSWORTH, AND
M. SCOTT HARRIS
2016   Coastal and Inland Geologic and Geomorphic
        Processes. In *Site Formation of Submerged
        Shipwrecks*, edited by Matthew E. Keith, pp. 17–43.
        University Press of Florida, Gainesville, FL.

GALVESTON DAILY NEWS
1903   Part of Cargo Taken. *Galveston Daily News* 30
        November, Galveston, TX.

HORLINGS, RACHEL L.
2011   Of His Bones are Coral Made: Submerged Cultural Resources, Site Formation Processes, and Multiple Scales of Interpretation in Coastal Ghana. Unpublished Ph.D. dissertation, Syracuse University, Syracuse, NY.

HORRELL, CHRISTOPHER
2005   Plying the Waters of Time: Maritime Archaeology and History of the Florida Gulf Coast. Unpublished Ph.D. dissertation, Florida State University, Tallahassee, FL.

JOHNSON, MALCOM L.
1966   "*Lake Austin*" A Coastal Trading Schooner. On file at the Texas Historical Commission, Marine Archeology Program.

KEITH, MATTHEW E.
2016   *Site Formation Processes of Submerged Shipwrecks.* University Press of Florida, Gainesville, FL.

MEIDE, CHUCK, JAMES MCCLEAN, AND EDWARD WISER
2001   *Dog Island Shipwreck Survey 1999: Report of Historical and Archaeological Investigations.* Research Reports No. 4. Program in Underwater Archaeology, Florida State University, Tallahassee, FL.

MORTON, ROBERT. A.
1994   Texas Barriers. In *Geology of Holocene Barrier Island Systems*, edited by R. A. Davis, pp. 75–114. Springer-Verlag, Berlin.

PAYNE, DARWIN
1970   Camp Life in the Army of Occupation: Corpus Christi, July 1845 to March 1846. *The Southwestern Historical Quarterly*, 73(3):326–342.

PEARSON, CHARLES E., AND JOE J. SIMMONS
1995   *Underwater Archaeology of the Wreck of the Steamship Mary (41NU252) and Assessment of Seven Anomalies, Corpus Christi Entrance Channel, Nueces County.* Report to United States Army Corps of Engineers, Galveston District, from Coastal Environments, Baton Rouge, LA.

REESE, PAULINE
1938   The History of Padre Island. Unpublished Master's thesis, Texas College of Arts and Industries, Kingsville, TX.

ROTH, DAVID
2010   *Texas Hurricane History*. National Weather Service, Camp Springs, MD.

RUSSELL, MATTHEW A.
2005   *Beached Shipwreck Archeology: Case Studies from Channel Islands National Park.* Submerged Resources Center Professional Reports, No. 18. Intermountain Region, National Park Service, Santa Fe, NM.

SCURLOCK, D.
1974   *Archeological Assessment: Padre Island National Seashore, Texas.* Office of the State Archeologist, Special Report, 1. Austin, TX.

SHEIRE, JAMES W.
1971   *Padre Island National Seashore: Historic Resource Study.* Office of History and Historic Architecture, Washington, D.C.

SMYLIE, VERNON
1964a   *The Secrets of Padre Island.* Texas News Syndicate Press.

1964b   *This is Padre Island, Land of Fantasy, Fun, Treasure and Adventure.* Texas News Syndicate Press, TX.

WEDDLE, ROBERT S.
1985   *Spanish Sea: The Gulf of Mexico in North American Discovery 1500-1685.* Texas A&M University Press, College Station, TX.

WEYMER, B.A., C. HOUSER, AND J. R. GIARDINO
2015   Post Storm Evolution of Beach-Dune Morphology National Seashore, Texas. *Journal of Coastal Research* 31(3):634–644.

WHITEHEAD, HUNTER W.
2021   *Reconnaissance Survey for the Bob Hall Pier Wreck (41KL108) Kleberg County, Texas.* Report to Nueces County Coastal Parks, Corpus Christi, TX. Prepared by Coastal Environments, Inc., Corpus Christi, TX.

. . . .

Hope Bridgeman
University of West Florida
11000 University Parkway
Pensacola, FL 32514-5750

Hunter W. Whitehead
420 Glenmore Street
Corpus Christi, Texas 78412

Site Formation Processes in the Mobile River: Analysis of Shipwreck Acoustic Imagery

Joseph Grinnan, Austin Burkhard

In 2018, SEARCH, Inc., archaeologists conducted archaeological investigations including a remote-sensing survey in the Mobile River near Twelve Mile Island, Mobile, Alabama. The survey resulted in the identification of 12 previously unknown shipwrecks and the relocation of another three previously known submerged cultural resources. Between July 2018 and November 2021, SEARCH, Inc., archaeologists collected updated acoustic imagery from six of these shipwrecks on up to four different occasions. Side-by-side comparison of collected imagery depicts substantial differences in both shipwreck characteristics and the adjacent environment that has occurred over a relatively short period of time. This paper reviews the site formation processes affecting shipwrecks within the Mobile River, including environmental setting and conditions. A discussion on the analysis of recorded acoustic imagery of these shipwreck sites also provides a means to better manage these resources in the future.

Em 2018, os arqueólogos da SEARCH, Inc. realizaram investigações arqueológicas, incluindo um levantamento por deteção remota no rio Mobile, perto de Twelve Mile Island, Mobile, Alabama. O estudo resultou na identificação de 12 naufrágios anteriormente desconhecidos e na relocalização de outros três recursos culturais submersos anteriormente conhecidos. Entre julho de 2018 e novembro de 2021, os arqueólogos da SEARCH, Inc. recolheram imagens acústicas atualizadas de seis destes naufrágios em quatro ocasiões diferentes. A comparação das imagens recolhidas mostra alterações substanciais tanto nas características dos naufrágios como no ambiente adjacente, que ocorreram num período relativamente curto. Este documento analisa os processos de formação que afetam os naufrágios no rio Mobile, incluindo as condições ambientais. Uma discussão sobre a análise das imagens acústicas registadas destes locais de naufrágio também fornece um meio para gerir melhor estes recursos no futuro.

En 2018, les archéologues de SEARCH, Inc., ont mené des recherches archéologiques, y compris un levé de télédétection dans la rivière Mobile près de l'île Twelve Mile, Mobile, Alabama. Le projet a permis d'identifier 12 épaves auparavant inconnues et la relocalisation de trois autres ressources culturelles submergées précédemment connues. Entre juillet 2018 et novembre 2021, les archéologues de SEARCH, Inc., ont recueilli des images acoustiques mises à jour de six de ces épaves et ce à quatre occasions différentes. La comparaison côte à côte des images recueillies illustre des différences importantes à la fois dans les caractéristiques des épaves et dans l'environnement adjacent qui sont survenues sur une période de temps relativement courte. Le présent document passe en revue les processus de formation des sites qui ont une incidence sur les épaves dans la rivière Mobile, y compris le cadre et les conditions environnementaux. Une discussion sur l'analyse des images acoustiques enregistrées de ces sites d'épaves fournit également un moyen de mieux gérer ces ressources à l'avenir.

Introduction

In 2018, SEARCH, Inc., archaeologists conducted archaeological investigations including a remote-sensing survey in the Mobile River near Twelve Mile Island, Mobile, Alabama. The survey resulted in the recordation of 12 previously unknown shipwrecks and the relocation of another three previously known submerged cultural resources. Between July 2018 and May 2022, a team of archaeologists under direction of the Alabama Historical Commission conducted multiple archaeological investigations in the Mobile River near Twelve Mile Island, Baldwin County, Alabama. To date, archaeologists

have collected acoustic imagery from six different shipwrecks on up to five separate mobilizations to monitor and assess site condition and site formation processes. Archaeological sites 1BA694, 1BA699, 1BA703, and 1BA706 were previously imaged in July 2018, while 1BA696/1BA702 and 1BA704 were previously surveyed in July 2018 and March 2020. The purpose of the August and November 2021 investigations was to assess the current condition of the six previously identified shipwrecks (sites 1BA694, 1BA699, 1BA696/1BA702, 1BA703, 1BA704, and 1BA706) and compare imagery to what was previously collected.

On 4 August 2021, SEARCH Inc., archaeologists conducted an underwater remote-sensing survey using a side-scan sonar to relocate and assess the condition of the archaeological sites 1BA694, 1BA699, 1BA696/1BA702, 1BA703, 1BA704, and 1BA706. The August 2021 survey resulted in the collection of approximately 1.9 line kilometers (km) (1.2 line miles [mi.]) of data collected along the southern shoreline. On 2 November 2021, archaeologists conducted an underwater remote-sensing survey using side-scan sonar to relocate and assess the condition of archaeological sites 1BA696/1BA702 and1BA704. The November 2021 survey resulted in the collection of approximately 1.0 line km (0.6 line mi.) of data. The surveys were designed to relocate the site, update imagery, and assess the condition of the previously identified shipwrecks. Although this paper reviews the site formation processes affecting shipwrecks within the Mobile River, the history and significance of these archaeological sites is important to fully understanding the resource. For a more in-depth discussion of Alabama's submerged cultural resources within the Mobile River the reader is directed to review the following reports: Delgado et al. (2018), Delgado et al. (2019a), Delgado et al. (2019b), Delgado et al. 2019c), Delgado et al. (2020a), Delgado et al. (2020b), and Marx et al. (2018).

Environmental Setting and Conditions

The July 2018 weather and environmental conditions during the remote-sensing and dive operations were favorable and sunny, with a light breeze. Air temperatures averaged 90 to 100 degrees Fahrenheit (F) (32–37 degrees Celsius [C]), with water temperatures averaging 90°F (32°C). Underwater visibility was extremely limited due to high amounts of sediment within the river. Tidal patterns consisted of high tide in the morning that gradually decreased as the day progressed. Water salinity near Twelve Mile Island is brackish and varies based on the tidal cycle. USGS 02470630, Mobile River at Barry Steam Plant NR Bucks located approximately 34 km (21 mi.) upstream of the survey area, recorded water levels at 1.1 meter (m) (3.5 feet [ft.]). Bottom composition within the Mobile River is silt and clay, with survey area water depths ranging from 0.9 m (3.0 ft.) to 6.0 m (20 ft.).

The February and March 2020 weather and environmental conditions during remote-sensing operations were favorable and sunny, with 0 to 5 knots sustained winds and flat seas. Air temperatures averaged 66 to 78°F (18–25°C), with water temperatures averaging 68°F (20 C). Tidal patterns consisted of high tide around 9:00 am. USGS 02470630, Mobile River at Barry Steam Plant NR Bucks located approximately 34 km (21 mi.) upstream, recorded water levels at 4.3 m (14.0 ft.). Bottom composition within the Mobile River is silt and clay, with survey area water depths ranging from shoreline to approximately 6.1 m (20.0 ft.).

Similarly. the August 2021 weather and environmental conditions during remote-sensing operations were favorable and sunny, with a 0 to 5 knots sustained winds and flat seas. Air temperatures averaged from 75 to 85°F (24 to 29°C), with water temperatures averaging 85°F (24°C). Tidal patterns consisted of high tide around 9:00 am. USGS 02470630, Mobile River at Barry Steam Plant NR Bucks located approximately 34 km (21 mi.) upstream, recorded water levels at 1.4 m (4.7 ft.). Bottom composition within the Mobile River is silt and clay, with survey area water depths ranging from shoreline to approximately 6.1 m (20.0 ft.).

Lastly, the November 2021 weather and environmental conditions during remote-sensing operations were favorable and sunny, with a 0 to 5 knots sustained winds and flat seas. Air temperatures averaged from 50 to 76°F (10 to 24°C), with water temperatures averaging 65°F (18°C). Tidal patterns consisted of high tide around 9:00 am. USGS 02470630, Mobile River at Barry Steam Plant NR Bucks located approximately 34 km (21 mi.) upstream, recorded water levels at 1.3 m (4.3 ft.). Bottom composition within the Mobile River is silt and clay, with survey area water depths ranging from shoreline to approximately 5.9 m (19.4 ft.).

Investigation of Sites 1BA694, 1BA699, 1BA696/1BA702, 1BA703, 1BA704, and 1BA706

In March 2018, archaeologists conducted a shallow-water investigation and diver evaluation of a semi-submerged shipwreck site in the Mobile River (Delgado et al. 2018). The March 2018 investigation resulted in the recordation of 1BA694 (Twelvemile Island Wreck) (Delgado et al. 2018). In July 2018, archaeologists conducted a Phase I remote-sensing survey within the Mobile River near Twelve Mile Island (Delgado et al. 2019a). The July 2018 survey resulted in the recordation of seven newly located shipwrecks, relocated four previously recorded archaeological sites, and identified numerous other potential submerged cultural resources (Delgado et al. 2019a).

Site 1BA694 (Twelvemile Island Wreck) was first located in March 2018 and first recorded in acoustic imagery collected in July 2018 as a 31.0 × 9.4 m (104.0 × 31.0 ft.) shipwreck. Diver and shallow-water investigation of site 1BA694 revealed the wreck to be a late-19-century or early-20-century schooner, likely built in the Pacific Northwest (Delgado et al. 2018). It may be the four-masted auxiliary schooner *Else*, built in the Pacific Northwest and abandoned in Mobile in 1928. Site 1BA699 (Target 001) was first recorded in acoustic imagery collected July 2018 as a 69.0 × 9.4 m (228.0 × 31.0 ft.) shipwreck. Diver investigation of site 1BA699 revealed the vessel to be an iron-hulled vessel of unknown type and historical context (Delgado et al. 2019a). Site 1BA696/1BA702 (Target 003) was first recorded in acoustic imagery collected July 2018 as a 72.0 × 7.9 m (237.0 × 26.0 ft.) iron- or steel-hulled shipwreck (Delgado et al. 2019a). It is likely the steamer *Lake Ellijay*, abandoned after partial scrapping in 1947. Site 1BA703 (Target 004) was first recorded in acoustic imagery collected July 2018 as a 24 × 10 m (81 × 36 ft.) shipwreck, with straight lines and 90-degree angles. Due to the barge-like characteristics of site 1BA703, diver investigation was not conducted (Delgado et al. 2019a). Site 1BA704 (Target 005) was first recorded in acoustic imagery in July 2018 as a 25.0 × 6.4 m (78.0 × 21.0 ft.) shipwreck. Diver investigations of site 1BA704 have identified the wreck *Clotilda* (Delgado et al. 2019b, 2019c). Site 1BA706 (Target 010) was first recorded in acoustic imagery collected July 2018 as a 28 × 7 m (95 × 23 ft.) shipwreck. Diver investigation of site 1BA706 revealed the vessel to be an iron-hulled vessel, with a raked bow similar to oceangoing sailing vessels (Delgado et al. 2019a, 2019c). It may be the iron-hulled sailing ship *Tusitala* (Delgado et al. 2020b).

In March 2020, archaeologists conducted a shallow-water investigation of site 1BA704 in the Mobile River (Delgado et al. 2020a). The purpose of this investigation was to assist the Alabama Historical Commission (AHC) as the legal owner and steward of the wreck with ongoing site management of the resource. Data acquisition and subsequent data processing provided AHC with an updated sonar record, which assists in documenting the site's exposed elements. The survey also served to document the condition of the wreck and the environmental factors potentially influencing the site's ongoing preservation. The March 2020 survey indicated that substantially more of site 1BA704 was exposed than previously observed in the March 2018 survey (Delgado et al. 2020a).

Remote-Sensing Field and Processing Methodology

Archaeologists conducted the August 2021 survey from an 8.5 m (28.0 ft.) aluminum-hull survey vessel and the November 2021 survey from a 6.4 m (21.0 ft.) aluminum, flat-bottom survey vessel. These vessels are ideally suited for the project location and environmental conditions and have ample deck space to conduct remote-sensing operations and are equipped with all the necessary safety equipment, including the appropriate number of life jackets, marine radio, horn, fire extinguisher, and visual distress signals. HYPACK, Inc., hydrographic navigation software, was used for vessel guidance. Instrumentation for the survey included a Trimble SPS356 differentially corrected global positioning system (dGPS) receiver, with a GA830 global navigation satellite system antenna and an EdgeTech 4125 dual-frequency (600/1,600 kilohertz [kHz]) CHIRP side-scan sonar.

The Trimble dGPS uses minimum shift keying (MSK) beacon or the Satellite Based Augmentation System to enhance the GPS positioning for improved, sub-meter-accurate real-time positioning. The 4125 side-scan sonar system uses CHIRP technology to provide higher-resolution imagery at ranges up to 50% greater than traditional continuous-wave systems operating at the same frequency. At 600 kHz, the 4125 can obtain resolution across tracks of 1.5 centimeters (cm) (0.6 inches [in]). The resolution improves to 0.6 cm (0.2 in.) at 1,600 kHz.

Archaeologists maintained consistent altitude of the side-scan towfish during the survey so that data acquisition met optimal survey standards. Side-scan sonar acoustics maintained an altitude above the bay floor between 10% and 20% of the selected range. This is achieved through a combination of instrument frequency and range, as well as towfish altitude. Archaeologists deployed the side-scan sonar towfish close to the vessel because water depths in the survey areas were not significant enough to adjust the altitude of these instruments. Vessel speed varied as well, but did not exceed 5 knots whenever possible, which maximized the data collection of each instrument.

HYPACK navigation software, interfaced with the dGPS, maintained vessel and equipment positioning, with sub-meter accuracy by means of layback calculations and logged real-time positional data. Archaeologists collected side-scan sonar imagery at a frequency of 1,600 kHz and a range of 30 m (98 ft.) (i.e., total swath width=60 m [197 ft.]). The dGPS was interfaced with the side-scan sonar topside acquisition computer operating EdgeTech Discover software, which embedded positional data into the raw imagery and allowed for geo-rectification of the side-scan sonar record during processing. Side-scan sonar was collected in a constant stream. The survey was conducted in the Alabama State Plane (west zone) coordinate system based on the NAD83 datum. All project data were incorporated into a geographic information system (GIS) geodatabase for organization, scientific analyses, and archiving.

Remote-Sensing Data Processing and Interpretation Methodology

Archaeologists reviewed raw side-scan sonar imagery to locate acoustic contacts indicative of cultural resources protruding above the bay floor. High-frequency imagery files (1,600 kHz) were imported into Chesapeake Technology, Inc., SonarWiz 7 sonar processing software utilizing settings adjusted for the EdgeTech 4125 acquisition methods. Following importation of the raw imagery, bottom tracking was performed to identify the first acoustic return, which determines the altitude of the towfish above the bay floor, creates a slant-range-corrected record, and removes the water column from the nadir region. Gain, color, and contrast settings were adjusted for each file to produce an optimal and even image across the entire mosaic. Individual shipwreck mosaic images were exported as geo-rectified images (geotiff format), with a resolution of 0.015-m/pixel (0.05-ft./pixel) and imported into ArcGIS 10.6 so that it could be layered with previously collected date and facilitate hazards analysis.

Remote-Sensing Results

As part of the assessment, archaeologists reviewed side-scan sonar data and generated scaled imagery for current and previous datasets. Remote-sensing data were processed following the methodology described above, and archaeologists applied the knowledge gained from previous survey work when interpreting the remote-sensing survey results.

Site 1BA694 (Twelvemile Island Wreck)

1BA694 is located in the Mobile River's eastern channel, across from Twelve Mile Island. In 2018, archaeologists recorded acoustic imagery of 1BA694 (Figure 1). In 2018 imagery, 1BA694 measured approximately 31.0

× 9.4 m (104.0 × 31.0 ft.). Acoustic imagery depicted a substantial amount of tree debris intermixed and built up around site 1BA694 (Delgado et al. 2019a).

In August 2021, archaeologists recorded updated acoustic imagery of site 1BA694 to assess overall site condition and review changes over time. Within 2021 imagery, site 1BA694 measures approximately 26 × 11 m (87 × 37 ft.) (Figure 1). The 2021 acoustic imagery depicts less tree debris than was observed in 2018; however, the shoreline vegetation near the starboard bow (eastern) has encroached and obfuscates imagery of the starboard structure. Furthermore, near midships and the western (downstream) portion of the shipwreck imagery depicts the likely accretion of substantial sediment, covering structural elements. It is possible that the western (downstream) portion of the shipwreck is degrading due to natural site formation processes, such as currents, tides, and storm events; however, diver investigation would be necessary to confirm overall sediment accumulation and timber integrity.

Site 1BA699

Site 1BA699 is located in the Mobile River's eastern channel, opposite Twelve Mile Island. In 2018, archaeologists recorded acoustic imagery of site 1BA699 (Figure 1). In 2018 imagery, site 1BA699 measures approximately 69.0 × 9.4 m (228.0 × 31.0 ft.). Acoustic imagery depicts a substantial amount of tree debris intermixed and built up around the site. Wooden pilings appear to extend vertically upward from the river bottom near the target. These timber features were visible protruding upwards from the water column (Delgado et al. 2019a).

In August 2021, archaeologists recorded additional acoustic imagery of site 1BA699 to assess overall site condition and review changes over time. Within 2021 imagery, site 1BA699 measures approximately 69.0 × 10.9 m (228.0 × 35.8 ft.) (Figure 1). A portion of the port stern exists within the nadir of 2021 imagery. Acoustic imagery depicts less tree debris near the site when compared to 2018 imagery. Additionally, accumulation of sediments on the channel side of the wreck, especially near the bow (eastern or upstream), may cover or obfuscate structural elements observed in 2018 imagery. It is possible that the eastern (upstream) portion of the shipwreck is degrading due to natural elements, such as currents, tides, and storm events; however, diver investigation would be necessary to confirm overall sediment accumulation and timber integrity.

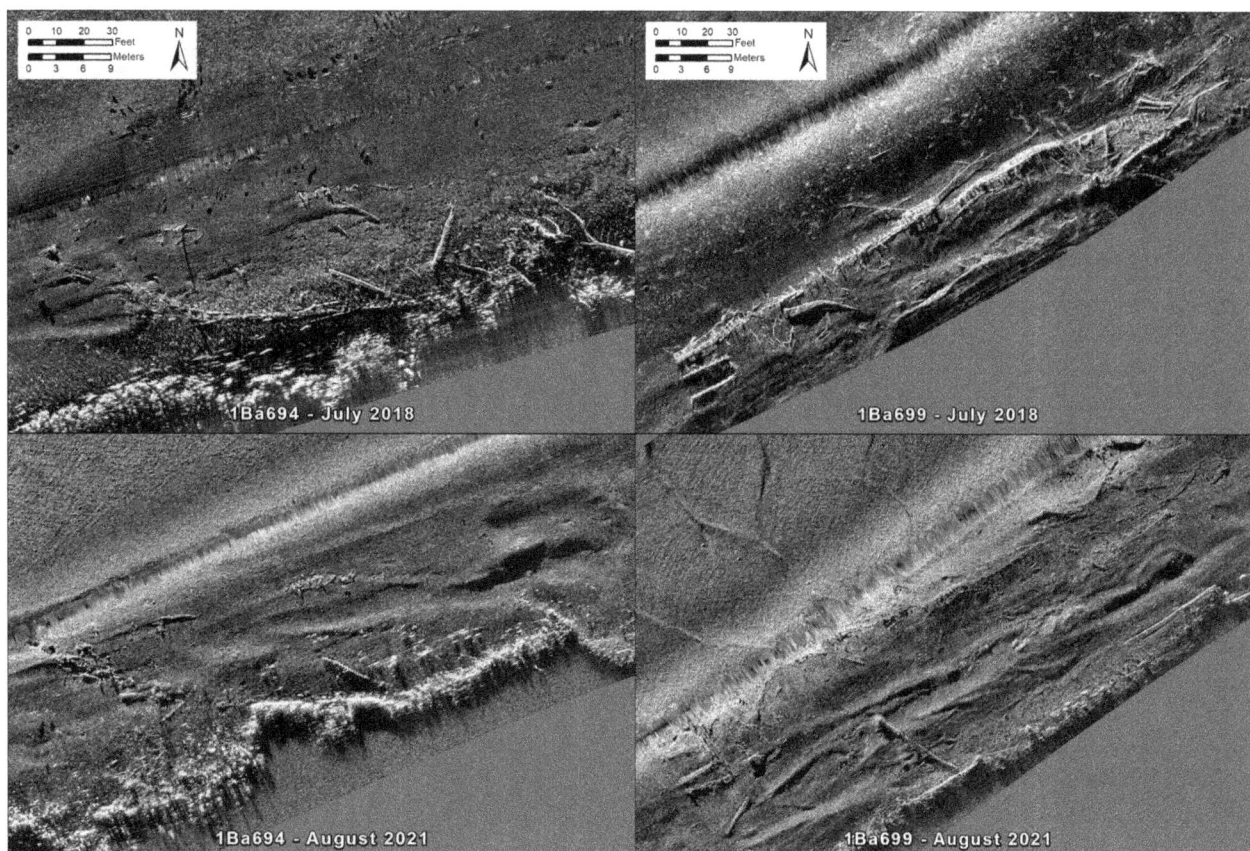

FIGURE 1. Site 1BA694 acoustic imagery, July 2018 (top left) and August 2021 (bottom left); Site 1BA699 acoustic imagery, July 2018 (top right) and August 2021 (bottom right). (Photo courtesy of the Alabama Historical Commission.)

Site 1BA696/1BA702 (SS Lake Ellijay*)*

Site 1BA696/1BA702 is located in Mobile River's eastern channel, along the bank opposite Twelve Mile Island, downriver of site 1BA699. In 2018, archaeologists recorded acoustic imagery of 1BA696/1BA702 (Figure 2). In 2018 imagery, 1BA696/1BA702 measures approximately 72.0 × 7.9 m (237.0 × 26.0 ft.). Acoustic imagery depicts tree debris around the hull of site 1BA696/1BA702. Portions of the bow are not clearly visible due to water levels and environmental conditions (Delgado et al. 2019a).

In August and November 2021, archaeologists recorded additional acoustic imagery of site 1BA696/1BA702 to assess overall site condition and review changes over time. Within 2021 imagery, site 1BA696/1BA702 measures approximately 70.0 × 7.9 m (229.0 × 26.0 ft.) (Figure 2). Acoustic imagery depicts tree debris near site 1BA696/1BA702 when compared to 2018 imagery. The 2020 imagery depicts similar imagery to that collected in 2018; however, 2021 imagery depicts substantial changes occurring between March 2020 and August 2021. An approximately 3.4 m (11.0 ft.) section of the aftmost portion of the stern is no longer extant

(Figure 2). Within 2020 imagery, this section contained articulated hull and decking elements, while within 2021 imagery, disarticulated frames and structural elements are visible along the river bottom. Additionally, approximately 11 m (36 ft.) forward of the stern the hull appears broken and shifted toward the river channel. This aligns with the forward portion of the aft hold, which may have been a weak point in the hull. Side-scan sonar shadowing along the starboard stern also depicts changes between 2020 and 2021 imagery. The side-scan sonar penetrates through the outer hull of the aft starboard side, indicating portions of the outer hull planking are no longer extant. Diver investigation would be necessary to investigate stern degradation and assess overall site integrity.

Site 1BA703

Site 1BA703 is located in Mobile River's eastern channel, along the bank opposite Twelve Mile Island, downriver of 1BA699. In 2018, archaeologists recorded acoustic imagery of site 1BA703, 15 m (50 ft.) from the riverbank, with approximate dimensions of 24 × 10 m (81 × 36 ft.) (Figure 3). The 2018 site condition may be

FIGURE 2. Site 1BA696/1BA702 acoustic imagery, July 2018 (top left), March 2020 (second from top left), August 2021 (second from bottom left), and November 2021 (bottom left); Site 1BA696/1BA702 acoustic imagery focused on stern section, July 2018 (top right), and November 2021 (bottom right). (Photo courtesy of the Alabama Historical Commission.)

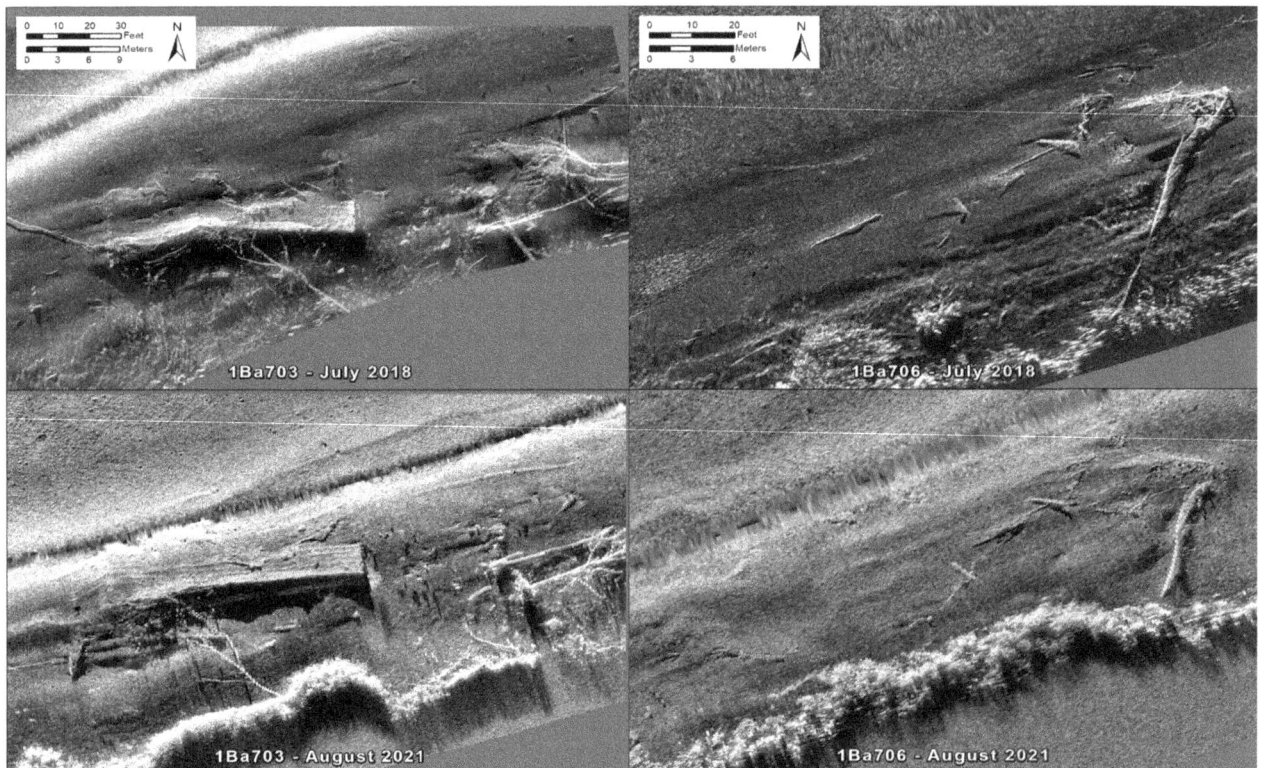

FIGURE 3. Site 1BA703 acoustic imagery, July 2018 (top left) and August 2021 (bottom left); Site 1BA706 acoustic imagery, July 2018 (top right) and August 2021 (bottom right). (Photo courtesy of the Alabama Historical Commission.)

serving to stabilize the riverbank from eroding into the center channel and is considered a hazardous environment (Delgado et al. 2019a).

In August 2021, archaeologists recorded updated acoustic imagery of site 1BA703 to assess overall site condition and review changes over time. Within 2021 imagery, site 1BA703 measures approximately 24 × 10 m (81 × 36 ft.) (Figure 3). The 2018 and 2021 acoustic imagery depict a substantial amount of tree debris intermixed and built up around the site; however, near midship and the western (downstream) portion of the shipwreck, a substantial amount of sediment has eroded, removing some tree debris, and exposing longitudinal ship elements. Based on the investigation of other nearby shipwrecks, these longitudinal lines may be supports at the intersection of iron decking, which are commonly found on iron-riveted barges. The decking between these supports has likely begun to sag since sinking, amplifying the supports visibility. Diver investigation would be necessary to investigate previously buried structural elements and assess integrity.

Site 1BAa706

Site 1BA706 is located in Mobile River's eastern channel, along the bank opposite Twelve Mile Island, downriver of site 1BA699. In 2018, archaeologists recorded acoustic imagery of site 1BA706 (Figure 3). In 2018 imagery, the site measures approximately 28 × 7 m (95 × 23 ft.). The 2018 imagery exhibits raised bow-like structure, hull-shaped outline, and depict a substantial amount of tree debris intermixed (Delgado et al. 2018, 2019c).

In August 2021, archaeologists recorded updated acoustic imagery of site 1BA706 to assess overall site condition and review changes over time. Within 2021 imagery, site 1BA706 measures approximately 23 × 12 m (77 × 40 ft.) (Figure 3). The 2018 and 2021 acoustic imagery depicts a substantial amount of tree debris intermixed and built up around the site; however, near midship and the bow (eastern or upstream) portion of the shipwreck, a substantial amount of sediment has accreted and may cover or obfuscate structural elements visible in 2018 imagery. It is possible that the bow is degrading due to natural elements, such as currents, tides, and storm events; however, diver investigation would be necessary to confirm overall sediment accumulation and timber integrity.

Site 1BA704 (Clotilda)

Site 1BA704 is located in Mobile River's eastern channel, along the bank opposite Twelve Mile Island, downriver of site 1BA699. In 2018, archaeologists recorded acoustic imagery of site 1BA704 along the riverbank with approximate dimensions of 25.0 × 10.9 m (78.0 × 36.0 ft.) (Figure 4). The 2018 site condition may be serving to stabilize the riverbank from eroding into the center channel and is considered a hazardous environment (Delgado et al. 2019a, 2019c). The 2018 investigation depicts the stern as mostly buried, with a substantial amount of tree debris intermixed and built up around site 1BA704. The March 2020 investigation indicated that the site was more exposed in 2020 imagery when compared to 2018 imagery, especially in the stern (Delgado et al. 2020a). The 2020 survey documented "additional features of the wreck and aspects of the site that speak to its integrity as an archaeological resource, while at the same time confirming post-sinking disturbances and impacts to the wreck" (Delgado et al. 2020a:18). The presence of an additional bulkhead in the main cargo hold to create two smaller spaces has been interpreted as reflecting modifications made to *Clotilda* for its final voyage to confine captives and segregate their food and water on its slave-trading voyage (Delgado et al. 2020b).

In August and November 2021, archaeologists recorded updated acoustic imagery of site 1BA704 to assess overall site condition and review changes over time. Within 2021 imagery, site 1BA704 measures approximately 26.0 × 7.9 m (88.0 × 26.0 ft.) (Figure 4). Acoustic imagery from 2018, 2020, and 2021 depicts a substantial amount of tree debris intermixed and built up around the site; however, the midship portion of shipwreck has less tree debris intermixed. Few changes were observed within acoustic imagery between 2020 and 2021. Additional diver investigation would be necessary to confirm overall site integrity. Additional investigation took place during the Phase III work in May 2022, with a summary report currently being prepared.

Conclusion

This discussion summarizes the results of two underwater remote-sensing surveys conducted in August and November 2021 in the Mobile River near Twelve Mile Island, Baldwin County, Alabama. The purpose of the surveys was to assess the condition of the six previously identified shipwrecks (sites 1BA694, 1BA699, 1BA696/1BA702, 1BA703, 1BA704, and 1BA706) and

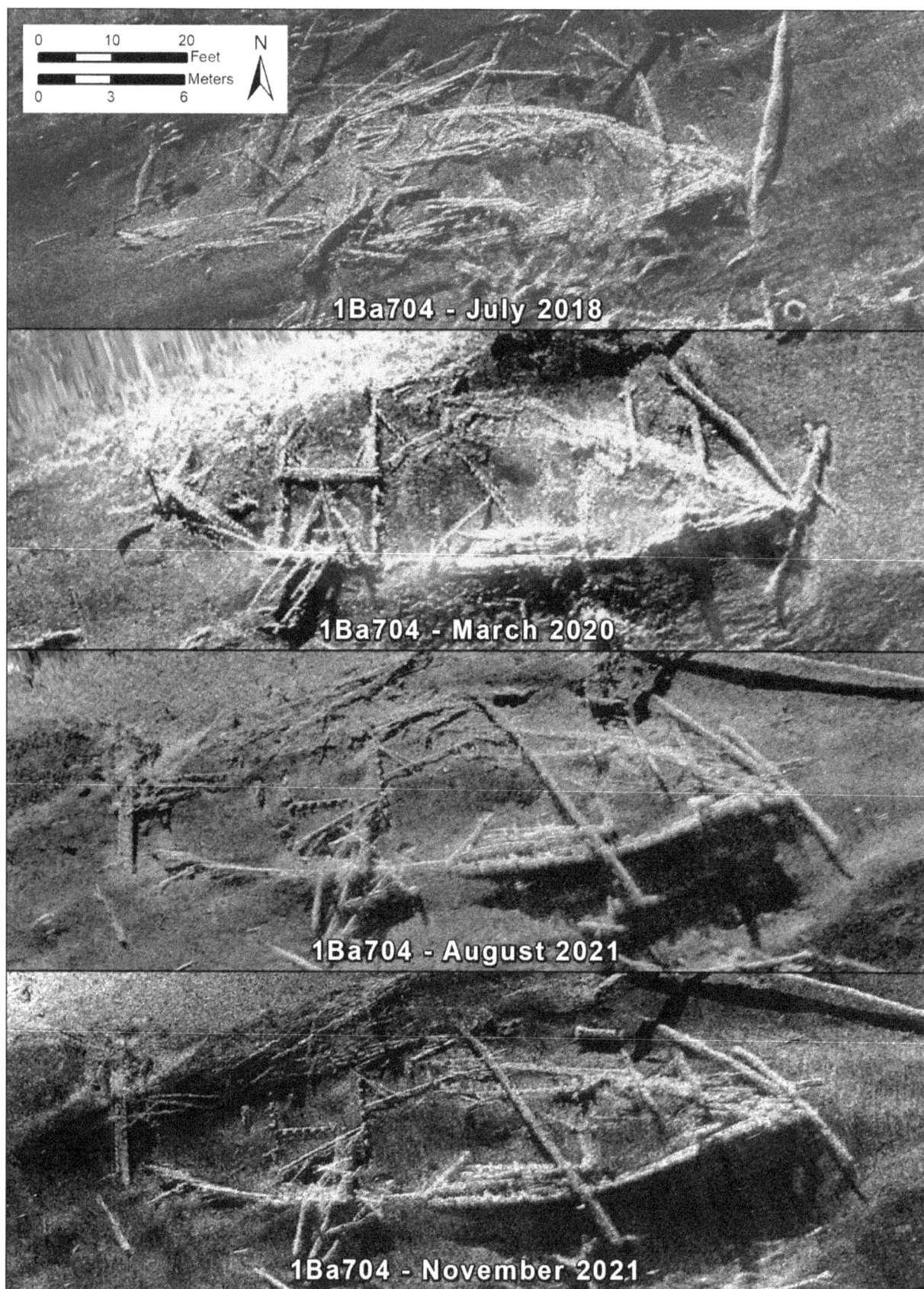

FIGURE 4. Site 1BA704 acoustic imagery, July 2018 (top), March 2020 (second from top), August 2021 (second from bottom), and November 2021 (bottom). (Photo courtesy of the Alabama Historical Commission.)

compare imagery to that collected in July 2018 as well as March 2020 (sites 1BA696/1BA702 and 1BA704, only).

Acoustic imagery depicts accumulation of sediments on sites 1BA694, 1BA699, and 1BA706, while imagery for 1BA696/1BA702, 1BA703, and 1BA704 provide evidence of sediment erosion and/or site changes, especially on the western (downstream) portion of each shipwreck. It is possible that the portion of sites 1BA694, 1BA699, and 1BA706 no longer visible on acoustic imagery have degraded or are no longer present due to natural elements, such as currents, tides, and storm events. Acoustic imagery from 2018 compared to 2020 depicts little to no change in site 1BA696/1BA702; however, 2021 imagery depicts substantial degradation of structural elements as having occurred since March 2020. Regarding site 1BA703, imagery from 2018 to 2021 depicts exposure of structural elements; however, no observable structural change is noted. Regarding site 1BA704, the 2018 to 2020 acoustic imagery depicts erosion/exposure of the stern. Comparison of 2020 and 2021 acoustic imagery between depict little to no obvious change on site 1BA704 following the more dramatic uncovering of the aft sections of the wreck.

Water levels during the 2021 survey were approximately 0.4 m (1.2 ft.) higher (water levels in August and November 2021 were similar) than they were during the 2018 survey and approximately 3.1 m (10.0 ft.) higher than during the 2020 survey. This depth difference and the resulting sonar towfish angle may account for some imagery differences. Additionally, water levels between July 2018 and August 2021 varied wildly and depict multiple flooding events, including a single spike of approximately 5.2 m (17.0 ft.) on the USGS gauge, or 4.0–4.3 m (13.0–14.0 ft.) more water than during data collection and eight additional spikes between 3.1 and 4.6 m (10.0 and 15.0 ft.) on the USGS gauge, or 1.8–3.4 m (6.0–11.0 ft.) higher than during data collection. These periods of flooding may account for, or contribute to, observed sediment and debris changes. Diver investigation would be necessary to verify sediment erosion or accretion, as well as individual timber and overall site integrity.

References

DELGADO, JAMES P., KYLE LENT, JOSEPH GRINNAN, D. MORGAN, STACYE HATHORN, AND DAVID CONLIN
2018   *Report on a Mid-to-Late Nineteenth-Century Wooden Shipwreck in the East Channel of the Mobile River Suggested as a Candidate for the 1855 Schooner Clotilda, Baldwin County, Alabama.* Report prepared by SEARCH, Inc., for the Alabama Historical Commission.

DELGADO, JAMES P., ALEX DECARO, JEFF ENRIGHT, JOSEPH GRINNAN, KYLE LENT, NICK LINVILLE, DEBORAH MARX, AND RAY TUBBY
2019a   *Mobile River Shipwrecks Survey, Baldwin and Mobile Counties.* Report prepared by SEARCH, Inc., for the Alabama Historical Commission.

DELGADO, JAMES P., DEBORAH MARX, KYLE LENT, JOSEPH GRINNAN, AND ALEX DECARO,
2019b   *Archaeological Investigations of 1Ba704, a Nineteenth Century Shipwreck Site in the Mobile River, Baldwin and Mobile Counties.* Report prepared by SEARCH, Inc., for the Alabama Historical Commission.

DELGADO, JAMES P., JOSEPH GRINNAN, ALEX DECARO, AND KYLE LENT
2019c   *Sector Scan Sonar Survey of 1Ba704 and 1Ba706 Baldwin County, Alabama.* Report prepared by SEARCH, Inc., for the Alabama Historical Commission and National Geographic Society.

DELGADO, JAMES P., ALEX DECARO, JOSEPH GRINNAN, KYLE LENT, AND MICHAEL BRENNAN
2020a   *Side-Scan Sonar Resurvey of 1Ba704, Baldwin County Alabama.* Report prepared by SEARCH, Inc., for the Alabama Historical Commission.

DELGADO, JAMES P., KYLE LENT, AND MICHAEL BRENNAN
2020b   *National Register of Historic Places Registration Form - Clotilda.* Nomination prepared by SEARCH, Inc., for the United States Department of the Interior, National Park Service.

MARX, DEBORAH, ALEXANDER DECARO, JOSEPH GRINNAN, KYLE LENT, RAY TUBBY, AND JEFFREY ENRIGHT.
2018   *Underwater Remote-Sensing Survey and Hazard Assessment of the Proposed Mobile River Barge Mooring Area, Baldwin County, Alabama.* Report Prepared by SEARCH, Inc., for the Alabama Department of Conservation and Natural Resources.

· · · · · · · · · · · · · · ·

Joseph Grinnan
1315 East Jordan Street
Pensacola, Florida, 32503

Austin Burkhard
1001 Simpson Street
Pensacola, Florida 32526

Assessing Northwest Florida's At-Risk Maritime Cultural Heritage Resources

Sorna Khakzad Knight, Barbara A. Clark

Northwest Florida encompasses unique cultural resources that have contributed to the development of North America and its history, supporting tourism, education, and recreation, which are important for Florida's socioeconomic development. At the state level, Florida is conducting an economic assessment of the impact of flooding on the state's real and natural resources. Focusing on the proposed National Heritage Area by Maritime Landscape Alliance for Preservation, this paper discusses the impact of flooding on cultural resources in Northwest Florida. It uses the same factors that the state has identified to assess other resources, along with information gathered from the Florida Public Archaeology Network, and the Florida Master Site File. The results highlight the sites that are at risk from flooding factors. The outcome of this research demonstrates the challenges in cultural preservation and public outreach, and in helping stakeholders prioritize actions for studying and preservation of cultural resources in Northwest Florida.

O Noroeste da Flórida engloba recursos culturais únicos que contribuíram para o desenvolvimento da América do Norte e da sua história, apoiando o turismo, a educação e atividades recreativas, que são importantes para o desenvolvimento socioeconómico da Flórida. A nível estatal, a Florida está a realizar uma avaliação económica do impacto das inundações nos recursos construídos e naturais do estado. Centrando-se na Área de Património Nacional proposta pela Maritime Landscape Alliance for Preservation, este documento discute o impacto das inundações nos recursos culturais do Noroeste da Florida. Utiliza os mesmos fatores que o estado identificou para avaliar outros recursos, juntamente com informações recolhidas da Florida Public Archaeology Network e do Florida Master Site File. Os resultados destacam os sítios que estão em risco devido a inundação. A investigação demonstra os desafios da preservação cultural e da sensibilização do público, e ajuda as partes interessadas a dar prioridade a ações de estudo e preservação dos recursos culturais no Noroeste da Florida.

Le nord-ouest de la Floride englobe des ressources culturelles uniques qui ont contribué au développement de l'Amérique du Nord et de son histoire, soutenant le tourisme, l'éducation et les loisirs, qui sont importants pour le développement socioéconomique de la Floride. Au niveau de l'État, la Floride mène une évaluation économique de l'impact des inondations sur les ressources réelles et naturelles de l'État. En se concentrant sur la zone du patrimoine national proposée par Maritime Landscape Alliance for Preservation, cet article traite de l'impact des inondations sur les ressources culturelles dans le nord-ouest de la Floride. Il utilise les mêmes facteurs que l'État a identifiés pour évaluer d'autres ressources, ainsi que des informations recueillies auprès du Florida Public Archaeology Network et du Florida Master Site File. Les résultats mettent en évidence les sites qui sont menacés par les facteurs d'inondation. Les résultats de cette recherche démontrent les défis de la préservation culturelle et de la sensibilisation du public, et ceux pour aider les parties prenantes à prioriser les actions pour l'étude et la préservation des ressources culturelles dans le nord-ouest de la Floride.

Introduction

Florida encompasses a rich cultural and archaeological history, including some of the oldest cities in America and numerous pre-colonial archaeological sites. Florida's geography and unique concentration of development within the first few feet above high tide make it especially vulnerable to extreme flooding. Additionally, Florida's coastal areas are prone to subsidence, and may be most affected by sea level rise and other flooding factors (Ericson et al. 2006). Furthermore, with climate change the probabilities of hurricane occurrence increase. Therefore, many of the state's cultural resources are exposed to coastal hazards. With sea-level rise (SLR), these resources will experience higher probability for inundation, damage from coastal erosion, and increased wave action during major storms.

The state is acting to assess and plan for future climate change and flooding impacts on its natural and real resources. The Office of Economic and Demographic Research (EDR) is a research arm of the Florida Legislature principally concerned with forecasting economic and social trends that affect policymaking, revenues, and appropriations. According to the 2021 Senate Bill 1954, Statewide Flooding and Sea Level Resilience:

> *"the Office of Economic and Demographic Research shall include in the assessment an analysis of future expenditures by federal, state, regional and local governments required to achieve the Legislature's intent of minimizing the adverse economic effects of inland and coastal flooding, thereby decreasing the likelihood of severe dislocations or disruptions in the economy and preserving the value of real and natural assets to the extent economically feasible. To the extent possible, the analysis must evaluate the cost of the resilience efforts necessary to address inland and coastal flooding associated with sea level rise, high tide events, storm surge, flash flooding, stormwater runoff, and increased annual precipitation over a 50-year planning horizon."* (CS/CS/SB 1954 2021)

This provides an opportunity for an inclusive assessment of different resources, including natural and cultural resources that significantly contribute to the economy and tourism development.

Background

The Florida Public Archaeology Network (FPAN) was established in 2005, consisting of eight regional centers throughout the state, to help stem the rapid deterioration of the state's buried past and to expand public interest in archaeology. The unique structure of FPAN presents the opportunity for statewide citizen engagement, while still allowing for variations to meet the needs of more localized communities. One focus area for FPAN is to identify the impact of climate change on the cultural and archaeological resources within the State of Florida. FPAN launched the Heritage Monitoring Scout (HMS) program in 2016 as a citizen science program, designed for monitoring archaeological sites impacted by climate change. It utilizes volunteers to assist in monitoring archaeological and historical sites. The Scouts are trained volunteers who conduct noninvasive documentation consisting of photographs and systematically filling out site evaluation forms. Each Scout also agrees to a code of ethics. The information they gather is entered into a FPAN-managed database and shared with the Florida

Division of Historical Resources and public land managers (FPAN 2023).

Between 2016 and 2018, FPAN conducted a project to assess cultural tourism in Northwest Florida. The result of that study revealed the wealth and significance of cultural, archaeological, and natural resources that this area encompasses. Following this study, a collaborative initiative was started by the University of West Florida (UWF) to establish Northwest Florida as a National Heritage Area (NHA). NHAs are regions that have nationally significant cultural and natural resources that tell a unique story about their area (NPS-DOI 2023a). By 2020, Maritime Landscape Alliance for Preservation (MLAP) was established to continue the NHA initiative. Along with moving forward to establish Northwest Florida as an NHA, MLAP also conducts studies and projects to help with educating the public, promoting preservation, and building resiliency for the future to protect cultural and natural resources. One of these projects aims to understand the impact of climate change on cultural resources. Apace with MLAP and FPAN, the State of Florida is conducting a study to estimate the impact of climate change and flooding on the state's real and natural resources. This paper aims to discuss different projects and studies to create a general assessment of cultural and archaeological resources in Northwest Florida. Goals include estimating the extent of climate change and flooding impact on the cultural resources in Northwest Florida, and to understanding the challenges, limits, and potential for creating better assessment strategies, which may result in planning for resiliency in the area, and protecting cultural resources for the future.

Purpose of Study

This paper attempts to gather information from the findings of both EDR and FPAN to assess cultural and archaeological resources that may be at risk due to SLR and storm surge within 50 years and up to the year 2100. Results of the first- and second-year findings from EDR are applied to scenarios they have selected for assessment of the impact of climate change and flooding, specifically on Florida's coastal cultural resources. The state of these resources has been compared with the reports from FPAN HMS to augment the reliability of the spatial analysis and to evaluate the challenges of the data collection by the volunteer monitors.

The ultimate goals of this paper are as follows:

- to identify risk zones and specific risks that threaten Florida's coastal cultural and natural resources;

- estimate the number of resources that are at immediate risk;

- and recommend actions for preservation of these resources, prioritization of action, and potential management strategies for the future.

Methodology

The first step in this analysis was to quantify the flood risk and the risk areas. Risk is defined as the combination of the probability of an occurring hazard and the exposure of resources to the hazard. For the purpose of this research, hazard is defined as flooding resulting from two flooding factors, including storm surge and SLR. These are the most probable flooding hazard factors for Florida (EDR 2022). For each of these flooding factors, multiple probability scenarios can be projected to define the hazard areas and to estimate resource exposure to these flood hazard factors. Exposure, for the purpose of this report, is defined as the number of cultural and archaeological assets within a hazard area. Two main steps were involved in developing this study, data collection and data analysis and modelling.

Data Collection

Data was compiled and used that had already been collected by reliable sources. No actual field study for data collection was conducted. The study uses three types of data.

a) *Data related to flooding and climate-change factors*: Through a literature review, this study identified possible sources of information from relevant federal and state agencies and organizations to benefit from the outcome of scientific and practical research that already exists. Relevant available information and data was compiled from federal sources, such as National Oceanic and Atmospheric administration (NOAA), United States Geological Survey (USGS), United States Army Corps of Engineers (USACE), and Federal Emergency Management Agency (FEMA), and state agencies, such as Water Management Districts and the Department of

Environmental Protection. The risk factors that were included in the analysis are SLR and storm surge.

b) *Data on cultural and archaeological resources*: The primary source is the Florida Master Site File. The Florida Master Site File is the State of Florida's official inventory of historical, cultural resources, archaeological sites, historical structures, historical cemeteries, historical bridges, historic districts, landscapes, and linear features (Florida Department of State 2023). The other source is the National Register of Historic Places, developed by the United States National Park Service, Department of the Interior (NPS-DOI), which maintains an ongoing inventory of cultural resources that have been evaluated as eligible for inclusion under criteria established by the National Historic Preservation Act (NHPA) (NPS-DOI 2023b).

c) *Data collected on archaeological sites assessment*: This data was gathered by the FPAN HMS program.

Data Analysis and Modeling the Impact of Flooding Factors on Cultural-Archaeological Resources

Modeling is one method to visually and analytically project the impact of factors and variables, in this case, climate factors on resources, such as cultural and archaeological sites. Two types of models were used for this study. The first one is based on the scientific studies that have already been done. This relied on the latest results from the USACE South Atlantic Coastal Studies (SACS) study that was released in February 2022 and NOAA SLR and storm surge models. The other one is ArcGIS modeling and analysis that was performed through superimposing flooding data on archaeological and cultural data. In the latter modeling, the study considered different scenarios of SLR and storm surge to assess the state of cultural and archaeological sites. The assessment mainly highlights sites and locations that may be affected by one or multiple flooding factors.

Delineating the Study Area

This study focuses on Northwest Florida and the assessment of its coastal zone (Figure 1). EDR has categorized Florida into three impact zones based on areas' exposure to flooding factors, risk level, and geography and proximity to the coast and water. These categories include High Impact Zone, Intermediate Impact Zone, and Dispersed Impact Zone (EDR 2023). The High Impact Zone and Intermediate Impact Zones, which are coastal and riverine areas, have been applied as a boundary for the assessments in this paper. A primary reason for selecting these two zones is that these coastal and shoreline areas are dynamic, and prone to coastal activities, storms, SLR and flooding.

Current Climate Trends and Flooding in Florida

A main factor that has short- and long-term impact on temporary and permanent flooding is SLR. Therefore, SLR is of particular importance, especially for coastal areas, such as Florida. SLR can intensify storm surge and high tides, and also cause permanent submergence of many coastal areas. Additionally, Florida is prone to hurricanes and tropical storms, with a common threat of storm surge. Flooding from storm surge depends on many factors, such as the track, intensity, size, and forward speed of the hurricane, and the characteristics of the coastline where it comes ashore or passes nearby (McInnes et al. 2003). Studies show that climate change has contributed and will likely continue to contribute to

the intensification of storm events like hurricanes. As a result, more and stronger storm surges and flooding can be expected (Wong-Parodi and Garfin 2022; Emanuel 2020; McInnes et al. 2003). The following sections present an overview of two major flooding factors of SLR and storm surge that may cause severe and complicated inundation in Florida, and have been the focus of EDR flooding assessment for the past two years.

Sea-Level Rise in Florida

Based on recent regional observations in Florida, projections anticipate an increase in the acceleration of sea level rise (Sweet et al. 2017; Volkov et al. 2019; Intergovernmental Panel on Climate Change [IPCC] 2021, 2019, 2013). Some impacts of SLR are already visible in Florida (Emrich et al. 2017). As SLR also affects the socioeconomic landscape and causes loss of infrastructures and services (South Florida Regional Climate Change Compact [SFRCCC] 2020), cultural

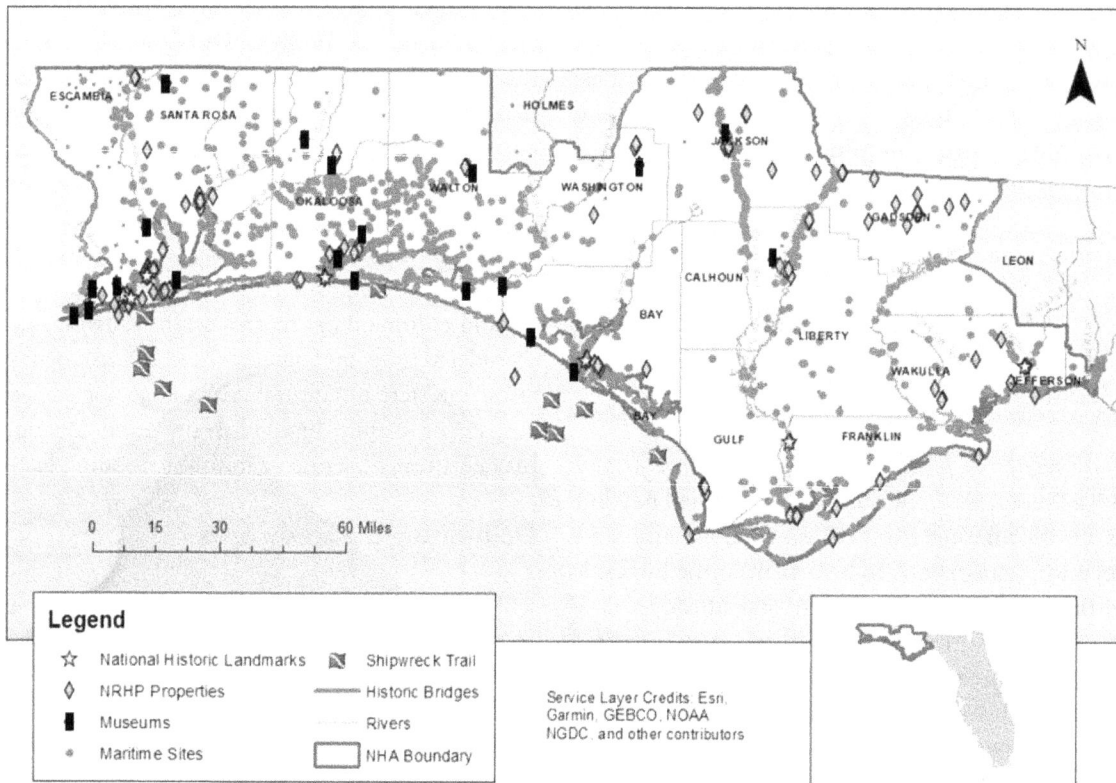

FIGURE 1. The Maritime Landscape Alliance for Preservation proposed National Heritage Area and some of its cultural and archaeological assets. (Map courtesy of Rachel Hines, produced for the Florida Panhandle Maritime Cultural Landscape National Heritage Area Feasibility Study, 2019.)

resources, including archaeological sites, historic buildings, and districts may negatively be affected, as well.

Scientific studies discuss and offer multiple scenarios of sea-level rise based on a variety of climate and environmental factors. NOAA released a new report and accompanying datasets from the U.S. According to this report, the 2005–2060 SLR projections show an average of 2 foot (ft.) SLR can be expected in the next 50 years in Florida (Sweet et al. 2022). EDR has selected a 2 ft. SLR as the most probable scenario for the next 50 years. The 2 ft. SLR scenario was also applied for this assessment to coastal cultural resources.

Storm Surge in Florida

Storm surge is an abnormal rise of water generated by a storm, over and above the predicted astronomical tides (Salmun and Molod 2015). Florida is already suffering from intense storm surges during hurricanes and storms (Parkinson et al. 2015). As a result of increasing sea levels, Florida has already experienced frequent loss-causing flood and wind events due to storm surge (Emrich et al. 2017), and probably will become more vulnerable to coastal flooding and storm surges in the near future (Florida Oceans and Coastal Council 2009).

Based on the NOAA Hurricane Research Division, from 1851 to 2020, 37 hurricanes hit different areas in Florida (NOAA 2022). Twenty-two of these hurricanes were identified as Category 3. Since 1995, nine hurricanes have hit Florida, of which five were Category 3. The remainder of hurricanes are categorized as Categories 4 and 5. As the state is experiencing hurricane Categories 4 and 5 in recent years, this paper has analyzed the impact of these two hurricane categories on the cultural and archaeological resources.

Assessment of Coastal Cultural Resources

Archaeologists from around the world are working to evaluate various aspects of climate change impacts on cultural and heritage resources (Anderson et al. 2017; Hambrecht and Rockman 2017; Rockman 2015; Rockman et al. 2016; Holtz et al. 2014). Projects, such as Scotland's Coastal Heritage at Risk Project (SCAPE); Changing Minds, Changing Coasts by the Coastal and Intertidal Zone Archaeological Network (CITiZAN); and the Climate, Heritage, and Environments of Reefs, Islands, and Headlands (CHERISH) have used archaeology to acquire data about how climate events are affecting these resources (Velentza 2022). In 2016, the FPAN developed the HMS program as a citizen science

initiative, inspired by the work of other similar programs, to document coastal and inland sites in Florida that are at risk from climate change and other anthropogenic factors (Miller and Murray 2018).

The proposed NHA area comprises 12,406 archaeological and cultural sites within its boundary. Most of the archaeological sites in this area consist of campsites, shell middens and mounds, artifact and lithic scatters, and quarries. Standing structures include historic houses, buildings, forts, lighthouses, schools, churches, and government buildings. Several historic landscapes and historic districts are also present. Many of the exposed archaeological sites are in the lowlands, deltas, and wetlands surrounding rivers and bays. Overlaying the Florida Master Site File data with the SLR and storm-surge scenarios shows that these factors affect the coastal areas differently. The low-lying areas in the eastern region of the panhandle, such as Apalachicola, and riverine areas may see more intense storm surge and flooding (Figure 2). Table 1 highlights the number of affected sites by four impacting factors.

As HMS data provides some information about the factors that are affecting the archaeological sites, analysis

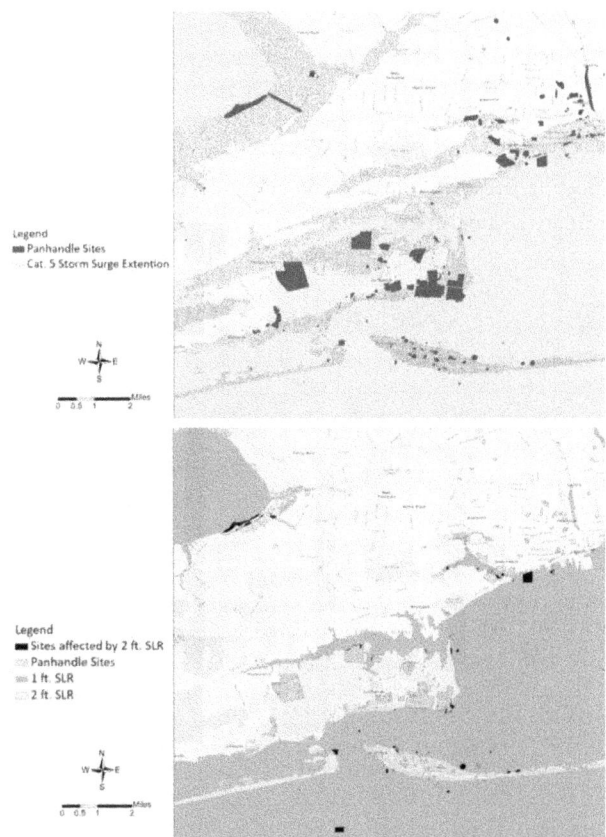

FIGURE 2. An example of sites overlaid with 2 and 3 ft. SLR scenarios, and hurricane category 5 storm surge. (Map by the primary author, 2023.)

was undertaken to understand the correlation between site locations and site status based on the HMS reports. Figure 3 shows a summary of this analysis. Accordingly, most of the sites close to the ocean and/or bays are reported as not found. Most of the sites that are close to a bay are reported in fair-to-declining state or not found. Some of the sites that are close to both bay and ocean are reported as poor-to-unstable, which was expected. However, against the working hypothesis, most sites

along the rivers or creeks are reported in a good-to-stable state (Figure 4).

To understand risk factors that are threatening sites, this research used HMS reports on archaeological and cultural sites. Ten risk factor categories were identified. These combined the visitor traffic with vehicle traffic, as sometimes no clear distinction was observed. Traffic was mentioned as the largest factor in disturbing, damaging, or destroying sites. Vegetation was also reported as a disturbance factor. However, it was not obvious from the data whether the vegetation caused damage to the sites/buildings or if it was an obstruction to assessing the sites. The risk factor reported as erosion is vague, since the erosion usually is a result of an element, such as wind, water, and so forth. In general, storms, waves, and flooding are the major degrading factors.

Discussion and Challenges

Flood is a complicated factor that unpredictably affects inland and coastal areas. Any projections are entirely hypothetical and based on assumptions that might not take into consideration all natural and anthropogenic factors. The aptitude of science and technology also plays an important role in making the analysis hypothetical at best. The uncertainties regarding SLR, changes in precipitation and magnitude of hurricanes and storm surges make it impossible to have a perfect assessment and prediction of future situations.

The impacts of hurricanes include a combination of strong winds, precipitation, and storm surge. Therefore, it is difficult to predict the exact impact of hurricanes due to variability in the strength of the wind, the amount of precipitation, and the duration of the storm. Presently, NOAA storm-surge data projects the extent of flooding caused by hurricane storm surge. The analysis, as such, cannot predict the amount of damage and destruction caused by wind or other factors, but the model offers an indication of the number of potentially impacted properties.

A challenge with archaeological site documentation is that many sites have not been monitored in regular intervals and some may not have been visited since their original discovery, which may have been decades ago. Therefore, changes due to SLR, erosion, looting, and other factors remain unknown, or not well recognized. Furthermore, the nature and location of some sites makes them challenging to access, which can inhibit regular monitoring. This makes it difficult to understand the direct and indirect impacts of climate change

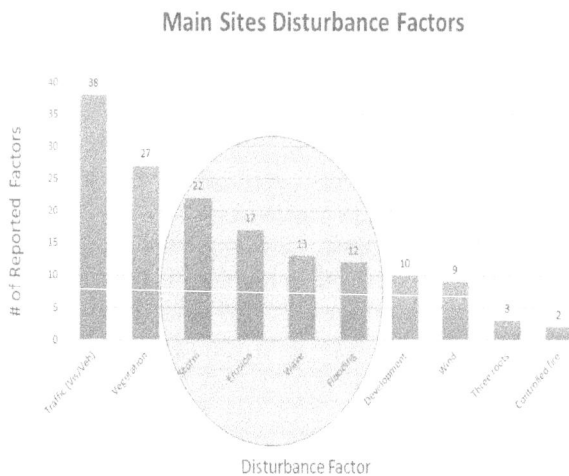

FIGURE 3. A summary of analysis, showing factors that are affecting the archaeological and cultural sites. (Graph by the primary author, 2023.)

FIGURE 4. The numbers of reported sites in relation to the sites' locations and their state of preservation. (Graph by the primary author, 2023.)

Factor Impacting Sites	2 ft. SLR	3 ft. SLR	Hurricane Cat. 5 Storm Surge	Hurricane Cat. 4 Storm Surge
Sites				
# Affected Florida Sites based on the authors' analysis (Total of 12406)	~996	~1154	3612	3179
# Affected HMS Sites reported by HMS Scout (Total of 133)	116	~116	Not reported by HMS	Not reported by HMS

TABLE 1. The number of affected sites due to four impacting factors.

on these resources, as well as within what time frame damage may be occurring. The HMS Program seeks to address this by gathering baseline data on both coastal and inland sites, making it possible to monitor changes from this point forward (Miller and Murray 2018). While numerous sites have been monitored throughout the panhandle region, thus far, it has not been systematic and no methodology for prioritizing sites based on SLR projections has been instituted. The data collected during this study can help to identify sites in need of monitoring, those that require subsequent monitoring, and the foundation for a method to prioritize sites that are projected to experience the most direct and dire impacts.

The HMS Program utilizes Arches, an open-source data management platform that was developed for cultural heritage management by the Getty Conservation Institute and World Monuments Fund. The platform is customizable and, because there is no licensing fee, it is financially accessible to most organizations. This platform allows for the collection and management of sensitive archaeological data due to its ability to create and customize different user permission settings. However, this proved to be a hindrance when it came to analyzing the data sets and comparing them to those that exist on other platforms. Without the ability to efficiently transfer and compare data, it had to be done by hand, which was time consuming and increased the potential for error. This is not seen as a weakness of Arches, however, as archaeological data is sensitive and protecting site data is an ethical and moral responsibility of any institution or individual who has access.

Spatial analysis of the impact of storm surge on archaeological and cultural resources would not provide a clear outcome of the impact. Since some sites are buried and some are exposed, the impact of flooding and waves varies significantly. Buildings, or remains of buildings, may also be affected differently, depending on the structures and materials. Middens and burial grounds could be affected in a variety of ways, depending on the state of exposure and other factors. In addition, the hurricane data sets only provide the extent of storm surge. This limited information cannot be used to project the impact of wind and rain on sites. Also, some sites are close to several bodies of water. These sites are at risk of more wave actions, storm surge, and high-tide events.

This study focused on the sites that were in the HMS study, with the purpose of having more information for analysis. The HMS program's goal is to provide baseline data for how climate change affects cultural resources, which also can be used as a tool for land managers to assess their sites and for disaster response (Miller and Murray 2018). This program also has the potential for use by policy makers to guide legislation and policy regarding the state's response to climate change. However, a few issues were raised during the analysis for this paper. First, why HMS chose certain sites and not others; are there any criteria to select these sites for the HMS Program? Second, the number of sites that are recorded in different locations does not follow any pattern. Third, the status of the assessed sites is subjective and depends on the volunteers' (Scouts') point of view. Fourth, the terminology that was used in the dataset was not consistent.

Following the EDR and FPAN projects and research, it is certain that a harmonized statewide action and recommendation plan can benefit the management of coastal resources. One example of a harmonized plan is the SFRCCC, a regional project in Southeast Florida, where four counties—Broward, Miami-Dade, Monroe, and Palm Beach—formed a climate compact in 2010 to address climate change impacts, including sea-level rise and high-tide flooding (SFRCCC 2020, 2017). To build resilience against climate change, the SFRCCC study concluded with some recommendations for different sectors, including, agriculture, energy and fuel, natural system, public health and so forth (SFRCCC 2020).

The SFRCCC work highlights the importance of regional coordination and collaboration. To improve regional coordination, the SFRCCC focuses on implementing a communication strategy, updating SLR projections, promoting broader project review, supporting resilience in jurisdictions, tacking regional indicators, and creating an equity group (SFRCCC 2022). Regional compacts, such as this, are an example of how successful climate action requires a coordinated regional approach and active engagement along with a change of attitude on the part of policymakers, stakeholders, and individuals (Velentza 2022).

Studying the impact of climate change on cultural resources has multiple benefits. The Climate Heritage Network released a manifesto in 2021 emphasizing the importance of cultural heritage as being both at risk due to climate change, but also part of the solution for mitigative action. The manifesto declared that people, their cultures, and both natural and cultural heritage are at risk due to climate inaction. This same document espouses the power of culture and heritage to inspire change and human action (Climate Heritage Network 2021). Although the connection between climate change and cultural heritage has been well noted, integrating archaeology into climate response has not been common practice (Van de Noort 2011). Archaeological science has a long tradition of studying past climate change and cultural adaptation to climatic events. This knowledge and understanding has the potential to contribute greatly to the study of and response to climate change, but has been underutilized.

Disclaimer: The views expressed in this paper are those of the authors and do not necessarily reflect the views or policies of the organizations that the authors affiliate with.

References

ANDERSON, DAVID G., THADDEUS G. BISSETT, YEKA J. STEPHEN, JOSHUA J. WELLS, ERIC C. KANSA, SARAH W. KANSA, KELSEY NOAK MYERS, R. CARL DEMUTH, AND DEVIN A. WHITE
2017   Sea-level Rise and Archaeological Site Destruction: An Example from the Southeastern United States using DINAA (Digital Index of North American Archaeology). *PLOS ONE* 12(11):e0188142.

CLIMATE HERITAGE NETWORK
2021   Accelerating Climate Action Through the Power of Arts, Culture, and Heritage: A Manifesto on Keeping 1.5 Degrees Alive. <https://agenda21culture.net/sites/default/files/manifesto_cultureatcop_en_final.pdf>. Accessed 17 April 2023.

COMMITTEE SUBSTITUTE FOR COMMITTEE SUBSTITUTE FOR SENATE BILL (CS/CS/SB) 1954
2021   CS/CS/SB 1954: Statewide Flooding and Sea Level Rise Resilience, in Laws of Florida, Chapter 2021-28, by the Committees on Appropriations; and Environment and Natural Resources, and Senators Rodrigues and Garcia. < http://laws.flrules.org/2021/28>. Accessed 24 May 2023.

ECONOMIC AND DEMOGRAPHIC RESEARCH (EDR)
2022   Annual Assessment of Flooding and Sea Level Rise, 2023 Edition, Chapter 6. <http://edr.state.fl.us/Content/natural-resources/2023_AnnualAssessmentFloodingandSeaLevelRise_Chapter6.pdf>. Accessed 17 April 2023.

EMRICH, CHRISTOPHER T., DANIEL P. MORATH, GREGG C. BOWSER, AND RACHEL REEVES
2017   Hazards and Vulnerability Research Institute, Climate-Sensitive Hazards in Florida Identifying and Prioritizing Threats to Build Resilience against Climate Effects. <https://flbrace.org/fl-vulnerability-assessment.html>. Accessed 17 April 2023.

ERICSON, P. JASON, CHARLES J. VÖRÖSMARTY, S. LAWRENCE DINGMAN, LARRY G. WARD, AND MICHEL MEYBECK
2006   Effective Sea-Level Rise and Deltas: Causes of Change and Human Dimension. *Global and Planetary Change* 50(1–2):63–82

FLORIDA DEPARTMENT OF STATE
2023   Master Site File, Florida Division of Historical Resources. <https://dos.myflorida.com/historical/about/division-faqs/master-site-file/#:~:text=Q%3A%20What%20is%20the%20Florida,historical%20bridges%20and%20historic%20district>. Accessed 06 February 2023.

FLORIDA OCEANS AND COASTAL COUNCIL
2009   The Effects of Climate Change on Florida's Ocean & Coastal Resources. Special report to the Florida Energy and Climate Commission and the people of Florida, Tallahassee, Florida. < https://floridadep.gov/rcp/rcp/content/reports-and-products>. Accessed 24 May 2023.

FLORIDA PUBLIC ARCHAEOLOGY NETWORK (FPAN)
2023   Heritage Monitoring Scouts, Florida Public Archaeology Network. <https://hms.fpan.us/>. Accessed 06 February 2023.

HAMBRECHT, GEORGE, AND MARCY ROCKMAN
2017   International Approaches to Climate Change and Cultural Heritage. *American Antiquity* 82(4):627–641.

HOLTZ DEBRA, ADAM MARKHAM, KATE CELL, AND BRENDA EKWURZEL
2014   National Landmarks at Risk: How Rising Seas, Floods, and Wildfires are Threatening the United States' Most Cherished Historic Sites. Union of Concerned Scientists. <https://www.ucsusa.org/sites/default/files/2019-09/National-Landmarks-at-Risk-Full-Report.pdf>. Accessed 06 February 2023.

INTERGOVERNMENTAL PANEL ON CLIMATE CHANGE (IPCC)
2013   *Climate Change 2013: The Physical Science Basis.* Contribution of Working Group I to the Fifth Assessment Report of the Intergovernmental Panel on Climate Change. Stocker, T.F., D. Qin, G.-K. Plattner, M. Tignor, S.K. Allen, J. Boschung, A. Nauels, Y. Xia, V. Bex and P.M. Midgley, editors. Cambridge University Press, Cambridge, England.

2019   Summary for Policymakers. In *IPCC Special Report on the Ocean and Cryosphere in a Changing Climate.* H.-O. Pörtner, D.C. Roberts, V. Masson-Delmotte, P. Zhai, M. Tignor, E. Poloczanska, K. Mintenbeck, A. Alegría, M. Nicolai, A. Okem, J. Petzold, B. Rama, N.M. Weyer, editors. Cambridge University Press, Cambridge, England.

2021   *Climate Change 2021: The Physical Science Basis. Contribution of Working Group I to the Sixth Assessment Report of the Intergovernmental Panel on Climate Change.* Masson-Delmotte, V., P. Zhai, A. Pirani, S.L. Connors, C. Péan, S. Berger, N. Caud, Y. Chen, L. Goldfarb, M.I. Gomis, M. Huang, K. Leitzell, E. Lonnoy, J.B.R. Matthews, T.K. Maycock, T. Waterfield, O. Yelekçi, R. Yu, and B. Zhou, editors. Cambridge University Press, Cambridge, England.

MCINNES, KATHLEEN L., KEVIN WALSH, GRAEME D. HUBBERT, AND T. BEER
2003   Impact of Sea-level Rise and Storm Surges in a Coastal Community. *Natural Hazards* 30:187–207.

MILLER, SARAH E., AND EMILY JANE MURRAY
2018   Heritage Monitoring Scouts: Engaging the Public to Monitor Sites at Risk Across Florida. *Conservation and Management of Archaeological Sites* 20(4):234–260.

NATIONAL OCEANIC AND ATMOSPHERIC ADMINISTRATION (NOAA)
2022   Continental United States Hurricanes. United States Hurricane Research Division. NOAA <https://www.aoml.noaa.gov/hrd/hurdat/UShurrs_detailed.html>. Accessed 2 March 2023.

NATIONAL PARK SERVICE, DEPARTMENT OF THE INTERIOR (NPS-DOI)
2023a   National Heritage Areas: Community-Led Conservation and Development. National Park Service, U.S. Department of the Interior. <https://www.nps.gov/subjects/heritageareas/index.htm>. Accessed 3 March 2023.

2023b   National Register of Historic Places. <https://www.nps.gov/subjects/nationalregister/index.htm>. Accessed 06 February 2023.

PARKINSON, RANDAL, PETER HARLEM, AND JOHN MEEDER
2015   Managing the Anthropocene Marine Transgression to the Year 2100 and beyond in the State of Florida U.S.A. *Climatic Change* 128(1):85–98.

ROCKMAN, MARCY, MARISSA MORGAN, SONYA ZIAJA, GEORGE HAMBRECHT, AND ALISON MEADOW
2016   *Cultural Resources Climate Change Strategy.* Washington, DC: Cultural Resources Partnerships and Science and Climate Change Response Program, National Park Service.

ROCKMAN, MARCY
2015   An NPS Framework for Addressing Climate Change with Cultural Resources. *SEMANTIC Scholar* 32(1):37–50.

SALMUN, HAYDEE, AND ANDREA MOLOD
2015   The Use of a Statistical Model of Storm Surge as a Bias Correction for Dynamical Surge Models and its Applicability along the U.S. East Coast. *Journal of Marine Science and Engineering* 3:73–86.

SOUTH FLORIDA REGIONAL CLIMATE CHANGE COMPACT (SFRCCC)
2017   Regional Climate Action Plan 2.0 [web tool]. South Florida Regional Climate Change Compact (SFRCCC), Broward, Miami-Dade, Monroe, and Palm Beach Counties, FL. <http://www.southeastfloridaclimatecompact.org/regional-climate-action-plan/>. Accessed 17 April 2023.

2020   *Unified Sea Level Rise Projection Southeast Florida.* A document prepared for the Southeast Florida Regional Climate Change Compact Climate Leadership Committee. p. 36.

SWEET, WILLIAM V., BENJAMIN D. HAMLINGTON, ROBERT E. KOPP, CHRISTOPHER P. WEAVER, PATRICK L. BARNARD, DAVID BEKAERT, WILLIAM BROOKS, MICHAEL CRAGHAN, GREGORY DUSEK, THOMAS FREDERIKSE, GREGRY GARNER, AYESHA S. GENZ, JOHN P. KRASTING, ERIC LAROUR, DOUG MARCY, JOHN J. MARRA, JAYANTHA OBEYSEKERA, MARK OSLER, MATTHEW PENDLETON, DANIEL ROMAN, LAUREN SCHMIED, WILL VEATCH, KATHELEEN D. WHITE, AND CASEY ZUZAK
2022   Global and Regional Sea Level Rise Scenarios for the United States: Updated Mean Projections and Extreme Water Level Probabilities along U.S. Coastlines. NOAA Technical Report NOS 01. National Oceanic and Atmospheric Administration, National Ocean Service, Silver Spring, MD, 111 pp. <https://oceanservice.noaa.gov/hazards/sealevelrise/noaa-nostechrpt01-global-regional-SLR-scenarios-US.pdf>. Accessed 17 April 2023.

Sweet, William V., Robert E. Kopp, Christopher P. Weaver, Jayantha Obeysekera, Radley M. Horton, E. Robert Thieler, and Chris Zervas
2017   Global and Regional Sea Level Rise Scenarios for the United States. Silver Spring, Maryland. <https://tidesandcurrents.noaa.gov/publications/techrpt83_Global_and_Regional_SLR_Scenarios_for_the_US_final.pdf>. Accessed 17 April 2023.

Van de Noort, Robert
2011   Conceptualizing Climate Change Archaeology. *Antiquity* 85(329):1039–1048.

Velentza, Katerina
2022   Integrating Maritime Archaeology and Maritime Cultural Heritage in the Pursuit of Sustainability and Climate Resilience. Ocean Decade Heritage Network, Denmark November 2022. <https://www.oceandecadeheritage.org/integrating-maritime-archaeology-and-maritime-cultural-heritage-in-the-pursuit-of-sustainability-and-climate-resilience/>. Accessed 17 April 2023.

Volkov, Denis L., Sang-Ki Lee, Ricardo Domingues, Hong Zhang, and Marlos Goes
2019   Interannual Sea Level Variability Along the Southeastern Seaboard of the United States in Relation to the Gyre-scale Heat Divergence in the North Atlantic. *Geophysical Research Letters* 46.13 (2019):7481–7490.

Wong-Parodi, Gabrielle, and Dana Rose Garfin
2022   Hurricane Adaptation Behaviors in Texas and Florida: Exploring the Roles of Negative Personal Experience and Subjective Attribution to Climate Change. *Environmental Research Letters* 17:034033.

· · · · · · · · · · · · · · ·

Sorna Khakzad Knight
Florida Legislature
Office of Economic and Demographic Research
111 W Madison Street, Suite 574
Tallahassee, Florida 32399

Barbara A. Clark
Florida Public Archaeology Network
3540 Thomasville Road
Tallahassee, Florida 32309

Determining National Register of Historic Places Eligibility of Artificial Reefs: A Hypothetical Case Study of Intentionally Sunk Ships and Other Objects off Pensacola, Florida

Hunter W. Whitehead

Artificial reefs are human-created structures, such as retired ships, barges, bridges, reef modules constructed of various materials, and other objects, which are placed underwater to promote marine life. The State of Florida's artificial reef program is one of the most active in the United States, with more than 3,800 public deployments within state and federal waters since the 1940s. The cultural significance of artificial reefs and their importance to the fishing and Self-Contained Underwater Breathing Apparatus community in Pensacola, Florida, are discussed here. Several reefs are examined to explore how National Register of Historic Places criteria for evaluation might be applied to determine National Register of Historic Places eligibility.

Os recifes artificiais são estruturas criadas pelo homem, tais como navios desativados, barcaças, pontes, módulos de recifes construídos com vários materiais e outros objetos, que são colocados debaixo de água para promover a vida marinha. O programa de recifes artificiais do Estado da Flórida é um dos mais ativos dos Estados Unidos, com mais de 3.800 instalações públicas em águas estatais e federais desde a década de 1940. O significado cultural dos recifes artificiais e a sua importância para a comunidade piscatória e de mergulho em Pensacola, Florida, são aqui discutidos. São examinados vários recifes para explorar a forma como os critérios de avaliação do National Register of Historic Places podem ser aplicados para determinar a elegibilidade para o National Register of Historic Places.

Les récifs artificiels sont des structures créées par l'homme, telles que des navires retirés, des barges, des ponts, des modules de récifs construits avec divers matériaux et d'autres objets, qui sont placés sous l'eau pour promouvoir la vie marine. Le programme de récifs artificiels de l'État de Floride est l'un des plus actifs aux États-Unis, avec plus de 3 800 déploiements publics dans les eaux étatiques et fédérales depuis les années 1940. L'importance culturelle des récifs artificiels et leur importance pour la pêche et la communauté des plongeurs en scaphandre autonome à Pensacola, en Floride, sont discutées ici. Plusieurs récifs sont examinés pour explorer comment les critères d'évaluation du Registre national des lieux historiques (National Register of Historic Places) pourraient être appliqués pour déterminer l'admissibilité à ce registre.

Introduction

Environmental historians and archaeologists have extensively studied historic and prehistoric uses of marine resources, the former to recognize ecological and socioeconomic trends (Bolster 2006), and the latter to understand coastal cultural adaptation (Stark and Voorhies 1978). According to a study by Marean et al. (2007), humans expanded their diets to include marine resources as early as ~164,000 years ago. Fish traps, or weirs, made of rocks and wooden posts have been documented along coasts around the world (O'Sullivan 2004); some of the earliest types of weirs date to at least the Mesolithic (Pedersen 1995). The practice of modifying marine and riverine environments as a foraging tactic persisted well into the modern era.

Instead of trapping fish swimming along streams or tidal lagoons, some cultures began constructing artificial reefs to encourage marine habitats. Some of the oldest documented installations of artificial reefs are bamboo structures used by Japanese fishing communities in the late 18th century (Ito 2011). In the United States, artificial reef construction occurred by the mid-19th century when Carolina fishermen found it necessary to sink log structures, with stones or live oak timber, to renew former fishing grounds (Holbrook 1860; Stone 1985). Lima et al. (2019) notes that some United States artificial reef activity occurred in the 1930s; however, the "real impetus for artificial reefs came from the observations of sunken vessels and planes of World War II in the 1940s" (Lima et al. 2019:81). Increased coastal recreational activities, such as sport fishing and Self-Contained Underwater Breathing Apparatus (SCUBA) diving led to widespread deployments of artificial reefs in the 1960s, using materials such as rocks, tires, rubble, wood, and even industrial materials, like oil platforms

(Baine 2001). Today, deployments of prefabricated reefs, retired ships, concrete bridge rubble, and other materials are typically permitted activities that benefit the recreational and commercial fishing and SCUBA-tourism industries.

The following examines a variety of artificial reefs offshore Pensacola, Florida, and attempts to evaluate several examples for a 'mock' determination of eligibility for the National Register of Historic Places (NRHP). Determining NRHP eligibility is key to the United States cultural resource regulatory framework, and is typically a triggered response to federal government undertakings that might impact historic or prehistoric resources. For context of the legal framework, see King (2008). Since many of these artificial reefs, with some exceptions, do not yet meet the requirement for a resource to be 50 years or older, a great deal of the discussion here is hypothetical. Much is deduced from Hardesty and Little (2009), and of course the *National Register Bulletin: How to Apply the National Register Criteria for Evaluation* (National Park Service, Department of the Interior [NPS-DOI] 1997). First, a short history of the Escambia County, Florida, artificial reef program is described below.

Escambia County Artificial Reef Program

Per Turpin (2009), artificial reefs have been intentionally constructed in the marine and estuarine waters of Escambia County, Florida, since the 1970s. With the intent to encourage fish population growth and increasing dive sites, commercial and charter fishing/SCUBA captains, organizations, municipalities, and individuals constructed artificial reefs with materials of opportunity. Automobile bodies, household appliances, boat hulls, and other items were placed within the waters of Pensacola Bay and the Gulf of Mexico, the locations of which were often "… kept secret to maintain control over the harvest of fishes from the artificial reefs" (Turpin 2009:2). Today, the Escambia County Marine Resources Division (ECMRD) oversees all permitting of artificial reefs in Pensacola waters, as required by the Florida Department of Environmental Protection and the United States Army Corps of Engineers. As of 2020, a total of 528 public artificial reefs are listed on the ECMRD Public Artificial Reef List (2020). The author is aware of several additional deployments since this list was updated. It should be noted that some listings, such as the freighter *San Pablo* and the battleship USS *Massachusetts* were intentionally sunk during military training operations.

How to Apply the National Register Criteria for Evaluating Artificial Reefs

As mentioned above, most artificial reefs offshore Pensacola were intentionally sunk after the early 1970s, and thus, based on age, are excluded from consideration for the National Register. However, some of the ships and barges were built and used from the 1940s onward, perhaps meeting the age requirement to be considered eligible. The question is raised as to whether a submerged ship is potentially eligible 50 years from the sinking event or from date of construction. Aside from age, for inclusion in the NRHP, a property or site must meet at least one of the following criteria:

- Criterion A: Be associated with important events that have contributed significantly to the broad pattern of our history.

- Criterion B: Be associated with the lives of persons significant in our past.

- Criterion C: Embody the distinctive characteristics of a type, period or method of construction; or represent the work of a master; or possess high artistic values; or represent a significant and distinguishable entity whose components may lack individual distinction.

- Criterion D: Have yielded, or may be likely to yield, information important in prehistory or history (NPS-DOI 1997).

The site or property must also have integrity of location, design, setting, materials, workmanship, feeling, and association. The *National Register Bulletin No. 20* (NPS-DOI 1992), which pertains to the nomination of historic vessels and shipwrecks, lists five basic vessels which may be eligible for listing in the NRHP: floating historic vessels, dry-berthed historic vessels, small craft, hulks, and shipwrecks. In this guidance, shipwrecks are defined as "a submerged or buried vessel that has foundered, stranded, or wrecked" (NPS-DOI 1992:3), ostensibly barring eligibility of intentionally sunk artificial reefs such as ships and barges. An argument for eligibility consideration for intentionally sunk ships, such as USS *Oriskany*, is that they serve as underwater museum ships and could be eligible if they meet one of the criteria.

Conversely, marine resources and the fishing industry, and their importance to the Pensacola community is illustrated in Grinnan's (2018) discussion of 19th- to 20th-century red snapper fishing. Given the longevity of the fishing industry, and its importance to the history of Pensacola, the author contends that artificial reefs are an extension of the industry's cultural significance to the region. The following case studies delve into the history of the aircraft carrier USS *Oriskany* and the Liberty ship *Joseph L. Meek*, and assesses the site significance of each vessel. Additional ships, oil platforms, World War-II tanks, and other materials sunk as reefs are also briefly considered.

Case Study: USS Oriskany

The first example is USS *Oriskany* (CV-34), which was intentionally sunk off the coast of Pensacola, Florida, in 2006 to become the world's largest artificial reef (Morgan et al. 2009). Built during the later years of World War II, *Oriskany* (Figure 1) was the last *Essex*-class ship built and was launched on 13 October 1945; however, the ship was decommissioned for modernization efforts until 1950 (Pawlowski 1971). *Oriskany* earned two battle stars for service in the Korean War, and ten during its campaigns in the Vietnam War. The Navy decommissioned the ship in 1976 and, eventually, sold it for scrap in 1995. In 1997, the Navy repossessed the ship and sold it to the State of Florida for the creation of an artificial reef in 2004 (Giberson 2011).

Prior to the 2006 deployment, cleaning and preparation of the future reef entailed removing hazardous materials, such as polychlorinated biphenyl (PCB). While the author is not certain what other materials may have been removed from the ship, in reference to National Register Bulletin No. 20, the preparation may certainly have affected the integrity of workmanship and feeling. According to the *Bulletin* (NPS-DOI 1992:9) integrity of workmanship "is maintained when materials are renewed in-kind," while integrity of feeling "means that the vessel evokes an aesthetic or historic sense of the past." Cutting of holes and removal of hatches on intentionally sunken vessels not only allows water in during the sinking process, but also provides a measure

FIGURE 1. USS *Oriskany* underway off the coast of Southern California, 27 January 1955. (Photo courtesy the Naval History and Heritage Command, Catalog # NH 97409.)

of safety via multiple options for divers to enter and exit the vessel.

At a location 22.5 nautical miles (mi.) south of Pensacola Pass, *Oriskany* currently sits upright in 212 feet (ft.) of water. From 2014 to 2019, the author served as a dive master aboard local charter vessels and witnessed slight changes in the durability of portions of the ship. Steel plates and other structure lie at the base of the superstructure, or island. The historical significance of the vessel is hardly in doubt; however, the vessel's integrity, structural and otherwise, combined with its intentional sinking as an artificial reef, hinder any future NRHP eligibility determinations.

Case Study: Liberty Ship Joseph L. Meek

The second example is the Liberty ship *Joseph L. Meek*, which was intentionally scuttled offshore Pensacola in 1976 as an effort by the United States government to dispose of the large number of laid-up Liberty-ship fleet vessels (Sawyer and Mitchell 1985:227). Liberty ships (Figure 2) were mass produced throughout shipyards across the United States in response to shipping needs of World War II. *Joseph L. Meek* was built in 1942 by the Oregon Ship Building Corporation in Portland, Oregon, and served primarily in the OWI James River Reserve Fleet (Sawyer and Mitchell 1985:123).

The United States Maritime Administrative Artificial Reef Program was established in 1972 under Public Law 92-402, which authorized the Secretary of Commerce to transfer obsolete Liberty ships to coastal states. According to Sawyer and Mitchell (1985), a total of 41 Liberty ships were scuttled through this program of nearly 3,000 ships built. Most others were either scrapped or were casualties of submarine warfare during World War II. Of note, SS *Jeremiah O'Brien* is one of only two remaining intact Liberty ships and serves as a museum ship in San Francisco, California (Jaffee 2008). The state of Texas received 14 Liberty ships through this program and, in 1998, Texas Parks and Wildlife published a bulletin describing their acquisition, the ships' histories, and the Texas Artificial Reef Program. While the bulletin primarily outlines the artificial reef program, it is interesting, the authors note, that the Texas Liberty ships are a good representative sample of the experience of the near 3,000 sister ships (Arnold et al. 1998:1).

Joseph L. Meek now lies 95 ft. below the water about 7 mi. offshore Pensacola Pass. Sawyer and Mitchell (1985) note that many Liberty ships destined for scuttling were

Figure 2. Example of a Liberty ship leaving the Delta Shipbuilding Co., New Orleans, Louisiana during World War II. (Photo courtesy the Naval History and Heritage Command, Catalog # 75822.)

dismantled to some degree, and while the author is not aware of any documentation regarding dismantling of this particular ship, the wreck consists of the hull of the ship and does not have a superstructure or decking. The dismantling of portions of the wreck likely deters future NRHP eligibility determinations. However, it is noteworthy that of the near 3,000 Liberty ships constructed in the United States only two remain intact, with a total of 41 submerged as artificial reefs, and approximately 200 lost to enemy action, weather, and accidents.

Ships, Oil Platforms, and Other Materials

The ECMRD list (2020) contains 28 other ships, tugs, dredges, oil rigs, and other materials repurposed for artificial reefs, most of which reached the end of their useful lives and were destined for the scrapyard. Some sites of note are the YDT 14 and YDT 15, navy dive tenders built in 1942 by Erie Concrete and Steel Supply Co., both of which were sunk in 2000. Deployed as an artificial reef in 1993, *Pete Tide II* is a 180-ft. steel vessel that was used to tow, anchor, and supply oil rigs. Two oil rig platform jackets, the Tenneco Rig, and the Chevron Rig, were deployed in 1982 and 1993, respectively, as part of the Rigs-to-Reefs Program. The Rigs-to-Reefs Program facilitated the conversion of obsolete rig platforms to artificial reefs, many of which were donated to coastal states (Dauterive 2000). Lastly, perhaps one of the earliest sanctioned artificial reefs in Pensacola waters, the Three Coal Barges, were being towed offshore in 1974 when they broke free during transport. United States Navy explosive experts sank them at their current location only a few miles off Pensacola Beach. An additional 27 sites are included on the list that consist of bridge rubble, pipes, culverts, pilings, and other construction materials. While it is difficult to establish whether the sunken ships, barges, and even rig platforms might eventually meet the NRHP criteria, it is unlikely the debris deposited offshore ever will receive consideration. Essentially trash, what may have previously been a bridge or other functioning structure, no longer retains any modicum of integrity.

Prefabricated Reef Modules

The majority, a total of 444 artificial reef sites documented on the ECMRD list, are prefabricated reef modules made from a combination of welded rebar, lime rock, and concrete in various designs, such as tetrahedrons, pyramids, reef balls, and other structures. Unlike materials of opportunity, by design, the prefabricated reef modules are constructed to be suitable substrates where algae and marine life may grow and consequently encourage fish population growth. Some of the reefs listed are flagged as no longer intact, missing, or buried. In the future, perhaps some archaeologist might consider these reef sites as part of the larger recreation fishing/SCUBA industry landscape; however, due to degrading materials, these structures may no longer exist.

Conclusion

A great deal of further consideration should be given to artificial reefs as the result of meaningful cultural practices. As humans continue to impact and utilize marine environments, future archaeologists may be able to glean the value that artificial reefs currently have to fishing and SCUBA communities. The author posits that the artificial reefs deployed in the recent past are as culturally significant as historic and prehistoric fishing weirs.

Acknowledgements

First, the author wishes to thank the Pensacola dive community for providing so much knowledge about the artificial reefs in the region for so many years. This paper would not be possible without time spent aboard the charter vessels of Dalton Kennedy, Douglas Hammock, and Andy Ross. Robert Turpin, Manager of the Escambia County Marine Resources Division, provided information of the local artificial reef program. Keith Mille with the Florida Fish and Wildlife Commission also provided information on Pensacola reef deployments and permitting.

References

Arnold, J. Barto, Jennifer L. Goloboy, Andrew W. Hall, Rebecca A. Hall, and J. Dale Shively
1998   *Texas' Liberty Ships: From World War II Working-Class Heroes to Artificial Reefs.* Texas Parks and Wildlife, Bulletin No. 99-1. Austin, TX.

Baine, Mark
2001   Artificial Reefs: A Review of their Design, Application, Management, and Performance. *Ocean Coast Management* 44:241–259.

Bolster, Jeffrey W.
2006   Opportunities in Marine Environmental History. *Environmental History* 11(3):567–597.

DAUTERIVE, LES
2000   Rigs-to-Reefs Policy, Progress, and Perspective.
        In *AAUS 20th Symposium Proceedings: Diving
        for Science in the Twenty-First Century*. American
        Academy of Underwater Sciences, pp. 64–66.

ESCAMBIA COUNTY MARINE RESOURCES DIVISION
(ECMRD)
2020   Public Artificial Reef List. Escambia County
        Marine Resources Division <https://myescambia.
        com/docs/default-source/upload/public-artificial-
        reef-list.pdf?sfvrsn=d1a9ef6c_2>. Accessed 7 March
        2023.

GIBERSON, ART
2011   *The Mighty O: USS Oriskany CVA-34.* Patriot
        Media Publishing, Niceville, FL.

GRINNAN, NICOLE BUCCHINO
2018   The Sailing Vessels of Pensacola's Historical
        Red Snapper Fishing Industry: Towards an
        Understanding of Construction Trends. *The
        International Journal of Nautical Archaeology*
        7(1):203–220.

HARDESTY, DONALD L., AND BARBARA J. LITTLE
2009   *Assessing Site Significance: A Guide for Archaeologists
        and Historians, 2nd edition.* Altamira Press,
        Lanham, MD.

HOLBROOK, JOHN EDWARDS
1860   *Ichthyology of South Carolina.* Russell and Jones,
        Charleston, SC.

ITO, YASUSHI
2011   Artificial Reef Function in Fishing Grounds off
        Japan. In *Artificial Reefs in Fisheries Management*,
        S.A. Bortone, F.P Brandi, G. Fabi, and S. Otake,
        editors, pp. 239–264. Boca Raton, FL.

JAFFEE, WALTER W.
2008   *The History of a Liberty Ship from the Battle of
        the Atlantic to the 21st Century, 2nd edition.* The
        Glencannon Press, CA.

KING, THOMAS F.
2008   *Cultural Resource Laws and Practice, 4th edition.*
        AltaMira Press, Lanham, MD.

LIMA, JULIANO SILVA, ILLANA ROSENTAL ZALMON, AND
MILTON LOVE
2019   Overview and Trends of Ecological and
        Socioeconomic Research on ArtificialReefs. *Marine
        Environmental Research* 145:81–96.

MAREAN, CURTIS W., MIRYAM BAR-MATHEWS, JOCEYLYN
BERNATCHEZ, ERICH FISHER, PAUL GOLDBERG, ANDY I.
R. HERRIES, ZENOBIA JACOBS, ANTONIETA JERARDINO,
PANAGIOTIS KARKANAS, TOM MINICHILLO, PETER J. NILSSEN,
ERIN THOMPSON, IAN WATTS, AND HOPE M. WILLIAMS
2007   Early Human Use of Marine Resources and
        Pigment in South Africa during the Middle
        Pleistocene. *Nature* 449:905–908.

MORGAN, O. ASHTON, D. MATTHEW MASSEY, AND WILLIAM
L. HUTH
2009   Diving Demand for Large Ship Artificial Reefs.
        Marine Resource Economics 24:43–59.

NATIONAL PARK SERVICE, DEPARTMENT OF THE INTERIOR
(NPS-DOI)
1992   *National Register Bulletin No. 20: Nominating
        Historic Vessels and Shipwrecks to the National
        Register of Historic Places.* National Park Service,
        Department of the Interior, Washington, DC.

1997   *National Register Bulletin: How to Apply the National
        Register Criteria for Evaluation.* National Park
        Service, Department of the Interior, Washington,
        DC.

O'SULLIVAN, AIDAN
2004   Place, Memory and Identity Among Estuarine
        Fishing Communities: Interpreting the Archeology
        of Early Medieval Fish Weirs. *World Archaeology*
        35(3):449–468.

PAWLOWSKI, GARETH L.
1971   *Flat-Tops and Fledglings: A History of American
        Aircraft Carriers.* A.S. Barnes and Company,
        London, UK.

PEDERSEN, L.
1995   7000 Years of Fishing: Stationary Fishing Structures
        in the Mesolithic and Afterwards. In *Man and Sea
        in the Mesolithic*, Anders Fischer, editor, pp. 75–86.
        Oxbow Books, Oxford, UK.

SAWYER, L.A., AND W.H. MITCHELL
1985   *The Liberty Ships: The History of 'Emergency' type
        Cargo Ships Constructed in the United States during
        the Second World War, 2nd edition.* Lloyd's of
        London Press, London, UK.

STARK, BARBARA L., AND BARBARA VOORHIES (EDITORS)
1978   *Prehistoric Coastal Adaptations: The Economy and
        Ecology of Maritime Middle America.* Academic
        Press, New York, NY.

STONE, RICHARD B.
1985   History of Artificial Reef Use in the United
        States. In *Artificial Reefs: Marine and Freshwater
        Applications*, Frank M. D'Itri, editor, pp. 1–9. CRC
        Press, Boca Raton, FL.

Turpin, Robert K.
2009   Escambia County Artificial Reef Plan (v.2009.1).
        Escambia County Marine Resources Division
        <https://myescambia.com/docs/default-source/
        sharepoint-natural-resources-management/
        Marine%20Resources/Personal%20
        Reefs/20090126esccoartreefplan.pdf?sfvrsn=11>.
        Accessed 7 March 2023.

· · · · · · · · · · · · · · · ·

Hunter W. Whitehead
420 Glenmore Street
Corpus Christi, Texas 78412

Next Generation of Explorers: Training Submerged Terrestrial Archaeologists

Amanda M. Evans, Ramie A. Gougeon

Interest in submerged landscapes has received greater attention in the last decade in large part because of the increasing availability of the technology required to access submerged archaeological sites. However, training in the technologies, analyses, and even contexts needed to discover and interpret submerged sites remains in a nascent stage of development. A recent National Oceanic and Atmospheric Administration grant included high-impact educational practices aimed at bridging this gap. A team of graduate students were offered mentorship and training in basic analytical techniques. Their new skills were applied to the processing of core samples taken from the western Gulf of Mexico and they will be contributing to a research paper on the results. Additionally, a course in submerged terrestrial archaeology was offered to professionals and students in August 2022. Alongside overviews of cultural historical contexts, and submerged site modeling and discovery, course participants discussed what a successful submerged terrestrial higher educational program might include.

O interesse pelas paisagens submersas tem recebido uma maior atenção na última década, em grande parte devido à crescente disponibilidade da tecnologia necessária para aceder a sítios arqueológicos submersos. No entanto, a formação nas tecnologias, análises e mesmo contextos necessários para descobrir e interpretar sítios submersos continua numa fase inicial de desenvolvimento. Uma recente bolsa da National Oceanic and Atmospheric Administration incluiu práticas educativas de grande impacto destinadas a colmatar esta lacuna. Foi oferecida a uma equipa de estudantes licenciados orientação e formação em técnicas analíticas básicas. As suas novas competências foram aplicadas ao processamento de amostras recolhidas na zona ocidental do Golfo do México e contribuirão para um trabalho de investigação sobre os resultados. Além disso, foi oferecido um curso de arqueologia terrestre submersa a profissionais e estudantes em agosto de 2022. Para além de uma visão geral dos contextos histórico-culturais e da modelação e descoberta de sítios submersos, os participantes no curso discutiram o que poderia incluir um programa de ensino superior de arqueologia terrestre submersa bem-sucedido.

L'intérêt pour les paysages submergés a reçu une plus grande attention au cours de la dernière décennie en grande partie en raison de la disponibilité croissante de la technologie nécessaire pour accéder aux sites archéologiques submergés. Cependant, la formation sur les technologies, les analyses et même les contextes nécessaires pour découvrir et interpréter les sites submergés reste à un stade naissant de développement. Une récente subvention de la National Oceanic and Atmospheric Administration comprenait des pratiques éducatives à fort impact visant à combler cet écart. Une équipe d'étudiants des cycles supérieurs s'est vu offrir du mentorat et de la formation sur les techniques analytiques de base. Leurs nouvelles compétences ont été appliquées au traitement d'échantillons de carottes prélevés dans l'ouest du golfe du Mexique et ils contribueront à article de recherche sur les résultats. De plus, un cours d'archéologie terrestre submergée a été offert aux professionnels et aux étudiants en août 2022. Parallèlement aux aperçus des contextes historiques culturels, et de la modélisation et de la découverte de sites submergés, les participants au cours ont discuté de ce qu'un programme d'enseignement supérieur terrestre submergé réussi pourrait inclure.

Introduction

Interest in submerged landscapes has received greater attention in the last decade in large part because of the increasing availability of the technology required to access submerged archaeological sites. However, training in the technologies, analyses, and even contexts needed to discover and interpret submerged sites remains in a nascent stage of development.

To begin to address this need, the public outreach elements of a National Oceanic and Atmospheric Administration, Office of Ocean Exploration and Research (NOAA OER) grant were designed to introduce students and professionals to several aspects of the study of submerged terrestrial landscapes.

Scope of Work

The project proposal was a direct response to NOAA OER FY (Fiscal Year) 2018, Theme 2, Archaeological Exploration of the United States exclusive economic zone (EEZ) and outer continental shelf of the Gulf of Mexico, which specified activities "identifying submerged, previously sub-aerial paleo-landscapes," and "providing initial archaeological and environmental characterization of … submerged archaeological sites" (NOAA 2017:7). Project goals included the creation of baseline characterizations of, and further delineation of archaeological horizons surrounding, the now-submerged paleolandscape associated with the approximate shoreline stand ca. 8,000 years Before Present (yrs B.P.) that would have been available to early human populations in the northwestern portion of the Gulf of Mexico. Fieldwork activities included geophysical survey and sediment coring in federal waters offshore Texas and Louisiana. The NOAA OER funding program includes requirements for outreach and/or educational opportunities that can help to train "the future generation of explorers" (NOAA 2017:20) and a range of opportunities were considered and implemented throughout the scope of work, which would include students or new professionals.

Geophysical Fieldwork

The geophysical survey was designed following a review of 238 offshore oil and gas industry survey reports in the region and identification of complementary geological research. The results of this review were used to identify survey areas, determine the order of priority survey areas, and inform the final line plans for data collection. Pre-existing data were used to identify the probable orientation of each incised valley feature, so that the primary survey lines could be oriented perpendicular to the valley axis, providing for the best resolution of the valley's cross section. Mobilization and transit time were calculated to develop a preliminary schedule and set realistic expectations for data collection. Given the relatively short project duration of 2 years, the work of reviewing pre-existing surveys and creating the survey plans was conducted by senior project personnel. On a smaller scale, or with a longer timeline, this type of work would be beneficial to students or new professionals.

Chirp subbottom profilers consistently provide reliable subsurface data in the survey areas, but this project presented an opportunity to collect both chirp and parametric subbottom data, which is being used with greater frequency due to its advertised ability to resolve near-surface features in high detail. Acquisition of both types of data along the same survey transects created an opportunity to directly compare and contrast the resulting subsurface profiles. Bathymetry data were also acquired, to allow for calculation of total depths below modern sea-level. Magnetometer data were acquired to identify potential hazards to subsequent coring operations presented by modern debris and infrastructure in the survey areas. As the survey was focused on features buried below the modern seafloor, sonar data were not acquired. The objectives of the survey included relocation and identification of previously interpreted high-probability landforms, exploratory survey along incised valley systems, and comparisons of the two primary survey instruments.

The survey data were obtained over a 12-day charter, with two of those days used for vessel and equipment mobilization, including fabrication of a new parametric sonar flange. Survey operations were conducted 24/7, with the final 2 days lost due to inclement weather. In total, the survey resulted in acquisition of over 650-line kilometers of data per sensor. The vessel included berth space for 4 scientists, including 3 senior staff and 1 new professional. The scientific staff were divided into 2 teams of 2, each monitoring survey data collection and directing adjustments to the survey plans in real-time.

Being onboard the vessel during mobilization and survey is an intensive educational opportunity, where participants learn firsthand the challenges of installing and networking the various sensors, and experience survey operations, such as equipment deployment and data acquisition. This opportunity is necessarily limited by the amount of physical space on the vessel for supplemental personnel and by the various insurance and/or safety training requirements, or certifications required for offshore personnel. As a result, unpaid volunteers are often not allowed on these vessels, as personnel must be covered by maritime liability insurance, which requires that they receive a salary for their time offshore. Safety requirements may include offshore medical clearance, water survival training, or industry-specific authorizations, which in the case of United States-flagged vessels may include a Transportation Worker Identification Credential (TWIC) issued by the United States Department of Homeland Security.

Following the conclusion of survey operations, the data were interpreted by senior project personnel. The interpreted data were used to collectively identify locations for coring and sampling of the buried landscapes that would be acquired during the next field season. Graduate students and new professionals were not involved in the data interpretation due to the high volume of data collected and the relatively short timeframe for interpretation and core selection. A significant barrier to more active participation in this project phase by students is access to the software required; individual licenses can cost several thousand dollars or more, and multiple different program licenses are needed to view, interpret, and manipulate the data. While new professionals should have access to all the necessary software through their employer, students' access will depend upon their individual institution's resources.

Geotechnical Fieldwork

A total of 30 core locations were selected based on a double-blind interpretation of the subbottom and parametric data, followed by joint review of the results and comparison of recommended core locations by the project team. A scientific team of two from the grant were onboard the vessel during coring operations, which were conducted during daylight hours only over a 5-day period. The grant personnel worked with the coring contractor to position the cores, examine and section the cores upon recovery, label the cores, and adjust planned core locations based on preliminary field results. A total of 39 cores were collected, each measuring approximately 3.0 inches (in.) (7.6 centimeters [cm]) in diameter, and up to 20.0-feet (ft.) (6.0 meters [m]) in length. Although the vessel had berth space for additional personnel, the grant personnel included one senior project team member and one new professional to comply with health and safety protocols enacted in response to the COVID-19 global pandemic. Following the cruise, the cores were taken directly to the coring company's lab and run through a multi-sensor core logger, while still sealed. Cores were then delivered to project personnel, who began core splitting and subsampling.

Training Opportunities

As previously stated, the OER grant program requires an outreach and education component; therefore, this project was designed to include funding to support graduate students through a partnership with faculty at the University of West Florida (UWF). The initial idea was to give as many graduate students as possible an opportunity to work on submerged landscapes, including

getting first-hand experience with core sampling as an educator-at-sea during coring operations, and by assisting with analysis. Ultimately, the at-sea component was negated by the global pandemic, and other more practical limitations (outlined above). Grant funding was used to support a single fully funded, one-year graduate assistantship, with a full tuition stipend for a graduate student for the 2022–2023 academic year, and a graduate seminar on shell analysis using samples from the collected cores. Another public outreach element was the development and delivery of a short course on submerged terrestrial landscapes open to undergraduate and graduate students, as well as new professionals. The course was offered to the public through the Continuing Studies program and did not require admission to, or enrollment in, UWF. Funding was built into the budget to support the continuing education short course, including off-setting tuition for all participants.

Details about selection and results of coring operations and subsequent analyses will be reported in final reports of activities to NOAA, conference papers, and publications in the months to come. Suffice to say, the onset of the global pandemic in early 2020 required an adjustment to the float plan. While the coring cruise was completed as planned, project partners in the United Kingdom were unable to participate and the total number of onboard project team members was restricted. Once the cores were obtained, lab closures and staffing limitations to allow social distancing created significant delays.

Travel restrictions and mandatory shifts to remote delivery of educational programs resulted in the continuing education course being delayed twice. The first attempt to build a class through UWF's Continuing Education Office was begun in fall 2019 but abandoned by spring 2021 as questions remained about potential lockdowns and the challenges of offering the desired course content in a remote-learning format. In spring 2022, coordination with speakers and advertising of the course was renewed. COVID-19 travel restrictions and other logistical issues required the adoption of a hybrid of live and virtual presentations to a live audience.

Core Sampling Project

The private and academic partnership represented by the project team resulted in the creation of opportunities that benefitted academic objectives, as well as grant outcomes. Increasingly, educators are seeking ways to offer "high impact practices" (HIP) to students. HIP experiences involve several elements, including giving students opportunities for meaningful interactions with peers and professionals during applications of knowledge in "real-world" experiences (Kuh et al. 2017). These experiences should take considerable time and effort, and be marked by expectations for high performance. Ultimately, results of HIP experiences should be shared publicly. With these ideas in mind, the authors developed a guided analytical activity for graduate students around the processing of some of the core samples collected during the survey.

Several cores were selected specifically for the presence of shells in one or more discrete lenses. The samples were pre-processed by Dr. Gus Costa at his lab in Texas, then delivered to Dr. Ramie Gougeon's lab at UWF. There, Gougeon provided lab space, geological sieves for size-sorting, a computer for data entry, and other material support, such as bags, brushes, hand lenses, and a microscope. Graduate students in Anthropology were sent an email requesting a statement of interest in the project and general availability for lab work. Respondents came nearly exclusively from a cohort of historical archaeology students, with interests in historical maritime archaeology. Among these self-selected students, a few have backgrounds in zooarchaeological coursework or other lab experiences in material culture. One had prior experience working on a submerged terrestrial archaeological site with Dr. Jessi Halligan at Florida State University. Another two are undertaking thesis projects related to submerged terrestrial archaeology.

Once the team was assembled, a remote training session with Costa was scheduled. Project co-Principal Investigator, Dr. Amanda Evans, gave an overview of the project and detailed how the core samples were collected. Costa then went over basic shell sorting procedures and general expectations about what the samples would yield. The session was recorded and stored on a shared drive for the students to access should any questions arise and for new team members joining the project at a later date.

The following outlines the general methodology. Each sample is size sorted using standardized geological sieves into 4.75 millimeters (mm), 2.00 mm, 1.00 mm, and >1.00 mm fractions. The sub-millimeter fractions are reserved for future analyses. Materials within the three largest size categories are classified by the students into whole shell valves (right/left), fragmented shell, weathered shell, bone, charcoal, macrobotanicals (like seeds or stems), miscellaneous plant matter, lithic debitage, sediment aggregates, and "other." After each class of materials is separated within each size fraction, counts and weights of each class of material are recorded in a spreadsheet. For the large shell (No. 4 sieve fraction), valves that have umbos present and are complete enough for side identification are sorted into 'left' and 'right' valves. These are then laid out on a tray with their contextual information and photographed. Pictures are also taken of any unusual items, concretions, or materials requiring a second opinion. All photographs are uploaded to a shared drive and labeled. Shell lengths—nearly exclusively on *Rangia* shell—are taken using digital calipers and entered in a spreadsheet.

Once work was underway, the students identified additional steps they wanted to take to improve workflow and standardize data collection. For example, they developed a tray tag to track when each class of size-sorted material from a particular sample had been sorted, photographed, and processed. They also established a group email to pose questions and make observations or suggestions to each other as they were often working independently.

The first team of five students logged approximately 36 hours processing over two dozen size-sorted fractions between late March and early May 2022. The data were shared with Costa, who ran some preliminary analyses. Costa also prepared additional samples for the team to process in fall semester 2022. Two of the students were able to continue their participation during the fall semester and logged additional hours on the project. As species of shellfish other than *Rangia* and oyster are present in the new samples, the two students were tasked with developing shell identification guidelines and adjusting the data collection spreadsheet.

Once the students have finished processing the current set of samples, they will be shown how these data can be used to characterize the environments at each core location. The students will help generate descriptions and interpretations of the data in a manuscript intended for publication in a journal where they will be included as coauthors.

Submerged Terrestrial Short Course

With the financial support of the NOAA OER grant, a course in submerged terrestrial archaeology was offered to professionals and students in August 2022. UWF's Continuing Education (CE) Office assisted Dr. Greg Cook and Gougeon with the internal logistics of offering CE credits and certificates of participation, creating an enrollment portal, coordinating financial issues, conducting post-course assessments, and advertising the course. The course was initially conceived as a two-week program of lectures and hands-on activities. As the challenges mounted over bringing experts and materials to UWF, working around summer field schools and presenter work schedules, as well as accomplishing the educational goals within the parameters of the budget, the program was reimagined as a 4-day course. This change was also made to reduce the financial burden of lodging for non-local participants, as well as the time commitment for those taking leave from work or other programs to attend. Changes in UWF's CE office policies during the development of the course resulted in all participants receiving CE credit.

Enrollees were solicited through social media, including via posts on UWF Department of Anthropology websites, associated Facebook pages, LinkedIn, and through email or other communications to related email listservs, universities, and educational organizations. The costs of the course were borne, in large part, by the grant and the generosity of the presenters. Tuition was paid for by the grant for five graduate student participants. Tuition for the other participants was also offset by the grant and a reduction in fees from the UWF CE Office.

Speakers were drawn from colleagues from cultural resource management (CRM), academia, and government. The specific content of the course evolved over the two years of preparation, but always centered on these main themes: cultural history, geophysical and environmental contexts, and current methodologies. Ultimately, a remarkable slate of presenters was assembled, covering such topics as colonizers and paleolandscapes, understanding geological and hydrological variables in determining site potential, modeling climatological and sea-level changes in the past and present, and several perspectives on surveying and sampling. Presentations on compliance-related issues in the United States, including Tribal engagement, and case studies from the United States, Australia, the North Sea, Canada, the Mediterranean, and beyond, were included.

At the end of the third day, course participants discussed what a successful submerged terrestrial higher educational program might include. Gougeon framed this discussion by pointing out that although UWF recognized the need for more training in submerged terrestrial archaeology, the university lacks some disciplines necessary to give students a comprehensive baseline understanding of geology, for instance. Clearly, UWF is not totally incapable of training students in maritime survey and cultural history—three presenters graduated

from UWF's undergraduate and graduate programs. The issue remains, however. How can any program help fill this need for more trained professionals given the range of skillsets, theories, and fundamentals employed in this pursuit? Topics identified during the 4-day course included technical, administrative, regulatory, archaeological, and "soft" skills drawing from a large number of disciplines. Presenters delved into Tribal consultation and public engagement, CRM "law" (or putting policies into practice), geoarchaeology, geographical information systems (GIS) and data management, project management, statistics, field training and fieldwork, and a suite of ever-changing software and hardware applications. While it is unlikely that any one academic program will be able to provide in-depth training in all these disciplines and skills, students and instructors discussed ways that CE, on-the-job training, and professional organizations can help archaeologists with professional development after graduation. This might include mentorships, such as those offered by the Geological Society of America (GSA), participating in multiple field schools at different institutions, and pursuing certifications through online courses and training.

Training the Next Generation of Scholars

Increased development on the world's continental shelves, coupled with a greater awareness of the need to engage with Tribal nations and traditional owners has resulted in a demand for archaeologists trained and experienced in the identification and study of submerged landscapes. The NOAA OER project, which was academically focused on the exploration of buried and submerged landscapes, may arguably have a more lasting impact through the educational opportunities it has delivered and will continue to provide to students and new professionals. Private and academic partnerships must work to help fill educational or experiential gaps for current students who are needed in today's work force, and can help steer the next generation towards the experience they will need to address maritime archaeology's needs in the future.

Acknowledgements

Funding for this project was provided through NOAA, OER Grant NA18OAR0110287. The data and opinions in this paper have not been formally disseminated by NOAA, and do not represent any agency determination, view, or policy.

References

Kuh, George D., Ken O'Donnell, and Carol Geary Schneider
2017   HIPs at Ten. Change: *The Magazine of Higher Learning* 49:16–8.

United States Department of Commerce, National Oceanic Atmospheric Administration (NOAA)
2017   Ocean Exploration FY 2018 Funding Opportunity. NOAA-OAR-OER-2017-2005296 Full Announcement. NOAA/National Oceanic and Atmospheric Administration, Washington, D.C.

· · · · · · · · · · · · · · · ·

Amanda M. Evans
Gray & Pape, Inc.
110 Avondale St.
Houston, Texas 77006

Ramie A. Gougeon
Department of Anthropology
University of West Florida
Building 13, Room 115
11000 University Parkway
Pensacola, Florida 32514

ACUA Award Winners for 2023

George Fischer Student Travel Award

The George R. Fischer Student Travel Award provides support in the sum of $1,000 (USD) for international students currently studying maritime archaeology to attend and present a paper at the annual Society for Historical and Underwater Archaeology conference. George Fischer, a founding member of the ACUA, long supported and advocated for student participation in the annual conference not only for the experience and to foster an exchange of ideas, but also for the opportunity to meet other students and professionals outside of their home countries. This award, in honor of George, supports the professional development of students embarking on their nascent careers. This, in turn, furthers the overall mission of the ACUA to foster the growth and development of underwater and maritime archaeology throughout our watery world.

Two students were recognized at the 2023 Conference in Lisbon:

Lindsay Wentzel, a Master's student at East Carolina University, Program in Maritime History, presented a poster on *New Investigations into the Radford Wreck: Interpreting a Candidate for Cape Lookout's Lost Whaler.*

Dominic Bush, a PhD candidate at East Carolina University, Integrated Coastal Sciences Program, presented a paper entitled, *Microbiologically-Influenced Corrosion of Submerged World War II Plane Wrecks: Case Studies from Hawai'i.*

FIGURE 1. Dominic Bush (award winner), Ashley Lemke (ACUA Ex-Officio), Lindsay Wentzel (award winner), and Julie M. Schablitsky (current SHA President) at SHA 2023 at award presentation.

ACUA & RECON Offshore Diversity, Equity, and Inclusion Student Travel Award

The ACUA & RECON Offshore Diversity, Equity, and Inclusion Student Travel Award provides support in the amount of $1,000 (USD) to a student currently studying maritime archaeology or a related field who is presenting a paper or poster on an underwater or maritime archaeology topic at the annual Society for Historical and Underwater Archaeology conference. The goal of this travel award is to increase diversity, equity, and inclusion and to encourage student involvement at the meetings. Diversity is inclusive of race, ethnicity, gender, sexual orientation, abilities, and socioeconomic background.

Megan Crutcher, a third year PhD candidate in the Nautical Archaeology Program at Texas A&M University received the 2023 award and presented a paper entitled *Appropriating Language: The Historical-Archaeological Context of 'Grumetes' In Sources on West African Mariners.*

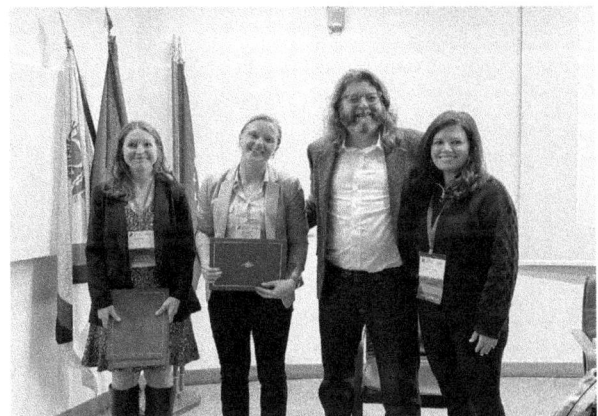

FIGURE 2. Ashley Lemke (ACUA Ex-Officio), Megan Crutcher (award winner), Michael C. Krivor (RECON Offshore), and Julie M. Schablitsky (current SHA President) at SHA 2023 award presentation.

ACUA Photo Competition Winners for 2023

If the old adage is true, "a picture is worth a thousand words," then the images created by archaeologists, historians, avocationals, and volunteers speak volumes about historical and underwater archaeology. Images can capture our imagination, take us to foreign lands, and show us faraway sites.

To honor our artistic colleagues, each year the ACUA sponsors a photo and video competition in conjunction with the annual Society for Historical Archaeology Conference on Historical and Underwater Archaeology. The competition is open to all SHA members and registered meeting attendees. The images are judged and displayed during the conference. Winners receive both a ribbon and the adulation of their peers.

Winning images from the 2023 photo competition are provided courtesy of the photographer and are presented below for all to enjoy. The winning video can be viewed on the ACUA YouTube channel available through the ACUA website (https://acuaonline.org/). Full color versions of the winning photographs are available on the ACUA website: (https://acuaonline.org/photo-contest/lisbon-2023/).

Category A: Color Archaeological Site

First: December 7th Survivor (YO-21) at Shipwreck Beach, Lana'i Hawai'i – Dominic Bush
Second: Inspecting a New Anchor in Biscayne – Stephanie Sterling
Third: The End of the Field Season – Sarah Noe
People's Choice: December 7th Survivor (YO-21) at Shipwreck Beach, Lana'i Hawai'I – Dominic Bush

Category B: Color Archaeological Field Work in Progress

First: Measurement of the propeller blades of the oil tanker Ilha Grande *(1962), in the Parcel de Manuel Luis Marine State Park, Maranhao, Brazil – Breatriz Bandiera*
Second: Moments in Time: 17th Century to Medieval Stratigraphy – Megan Olshefski
Third: Military veteran citizen scientists excavating the wreckage of a WWII aircraft, Saipan, CNMI
People's Choice: Measurement of the propeller blades of the oil tanker *Ilha Grande* (1962), in the Parcel de Manuel Luis Marine State Park, Maranhao, Brazil – Breatriz Bandiera

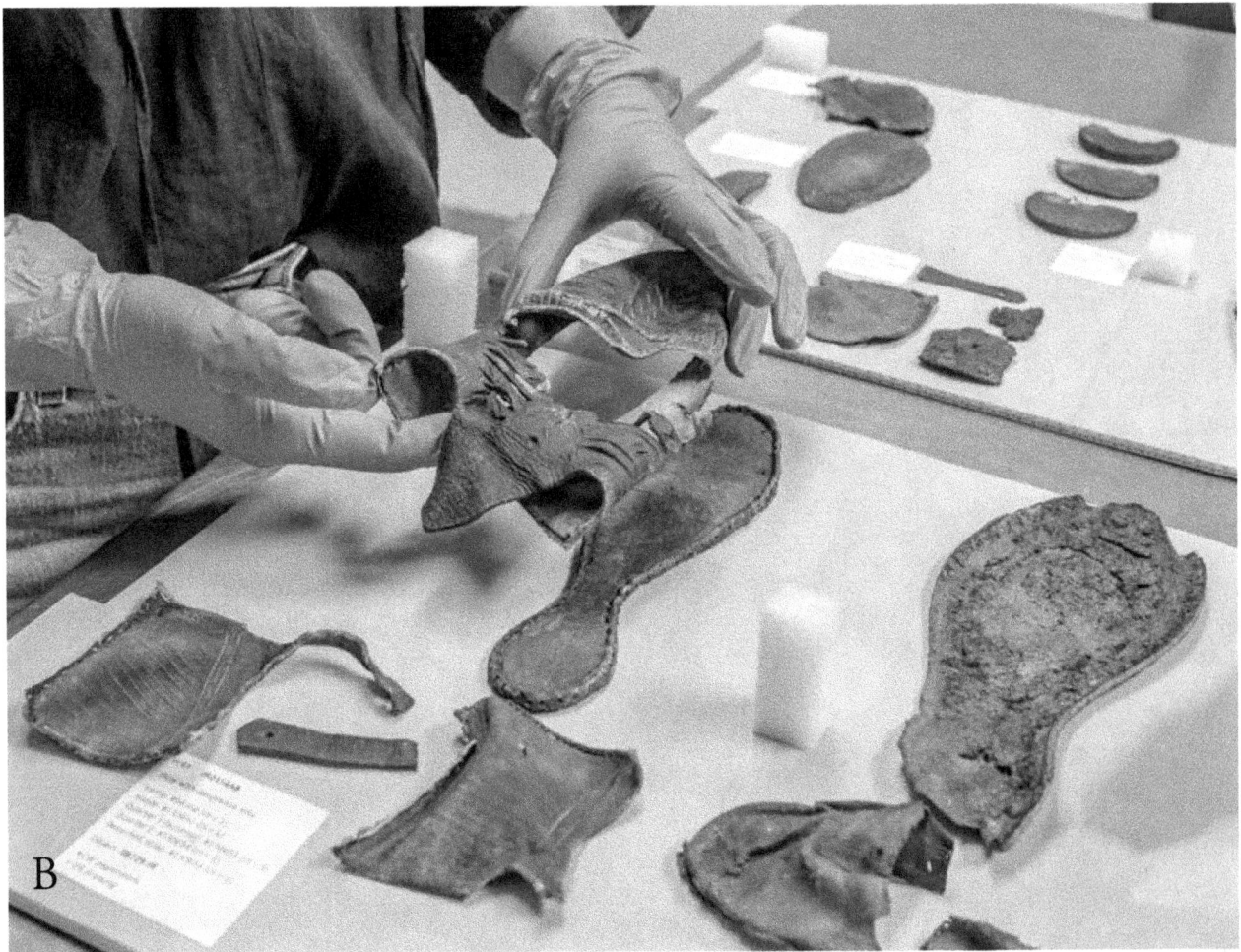

Category C: Color Archaeological Lab Work in Progress

First: Jamestown's Shoe Box: Finding Matching Sets – Chuck Durfor
People's Choice: Jamestown's Shoe Box: Finding Matching Sets – Chuck Durfor

(Category did not receive enough submissions for 2nd or 3rd place awards.)

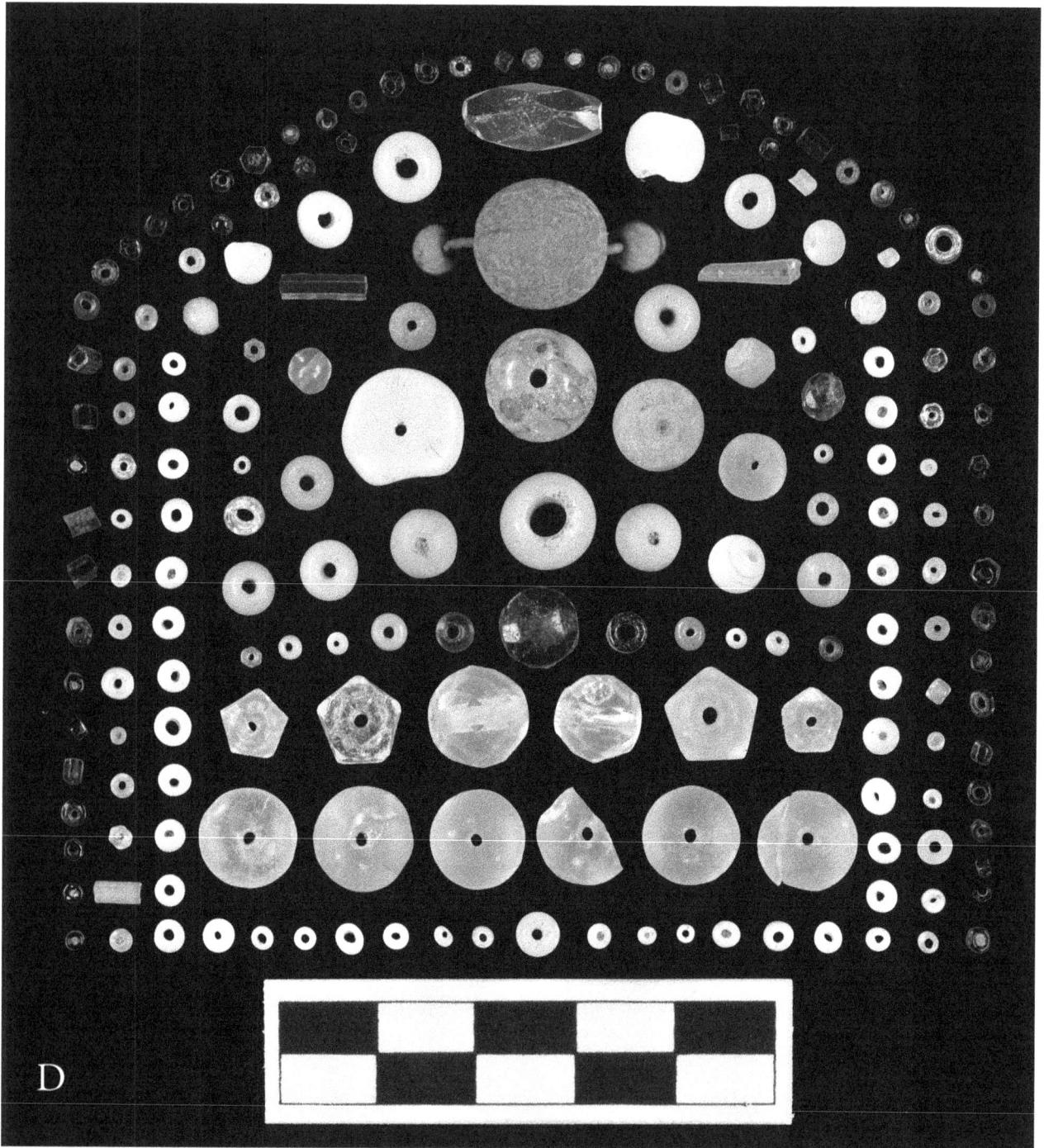

Category D: Color Artifact

First: Beads from the 1886-1939 Woodville School in Gloucester, Virginia – Colleen Betti
Second: Hidden Artistry of a French Pocket Watch – Emily McMillon
Third: Jolly Characters: Bartmann Jugs from Jamestown – Chuck Durfor & Leah Stricker
People's Choice: Beads from the 1886-1939 Woodville School in Gloucester, Virginia – Colleen Betti

E

Category E: Black & White Image

First: Taxing Textiles: James I Alnage Seal - Cathrine Davis

Second: High Contrast at the Wreckage of a Japanese Submarine Chaser, Saipan, CNMI – Nicole Grinnan

Third: YO-21 (WWII US Navy Oiler) at Rest on Lana'i's Shipwreck Beach - Dominic Bush

People's Choice: Taxing Textiles: James I Alnage Seal - Cathrine Davis

Category F: Color Archaeological Portrait

First: Simon Brown on the Dunraven Shipwreck - Alicia Johnson
Second: Student Examining 16th Century Customs Tag - Sarah Muckerheide
Third: Finding an Early 17th Century Well - Gabriel Brown
People's Choice: Simon Brown on the Dunraven Shipwreck - Alicia Johnson

Category G: Diversity

First: Bliss in Biscayne – Stephanie Sterling
Second: Egyptian Archaeologist, Asmaa Elsayed – Alicia Johnson
People's Choice: Egyptian Archaeologist, Asmaa Elsayed – Alicia Johnson

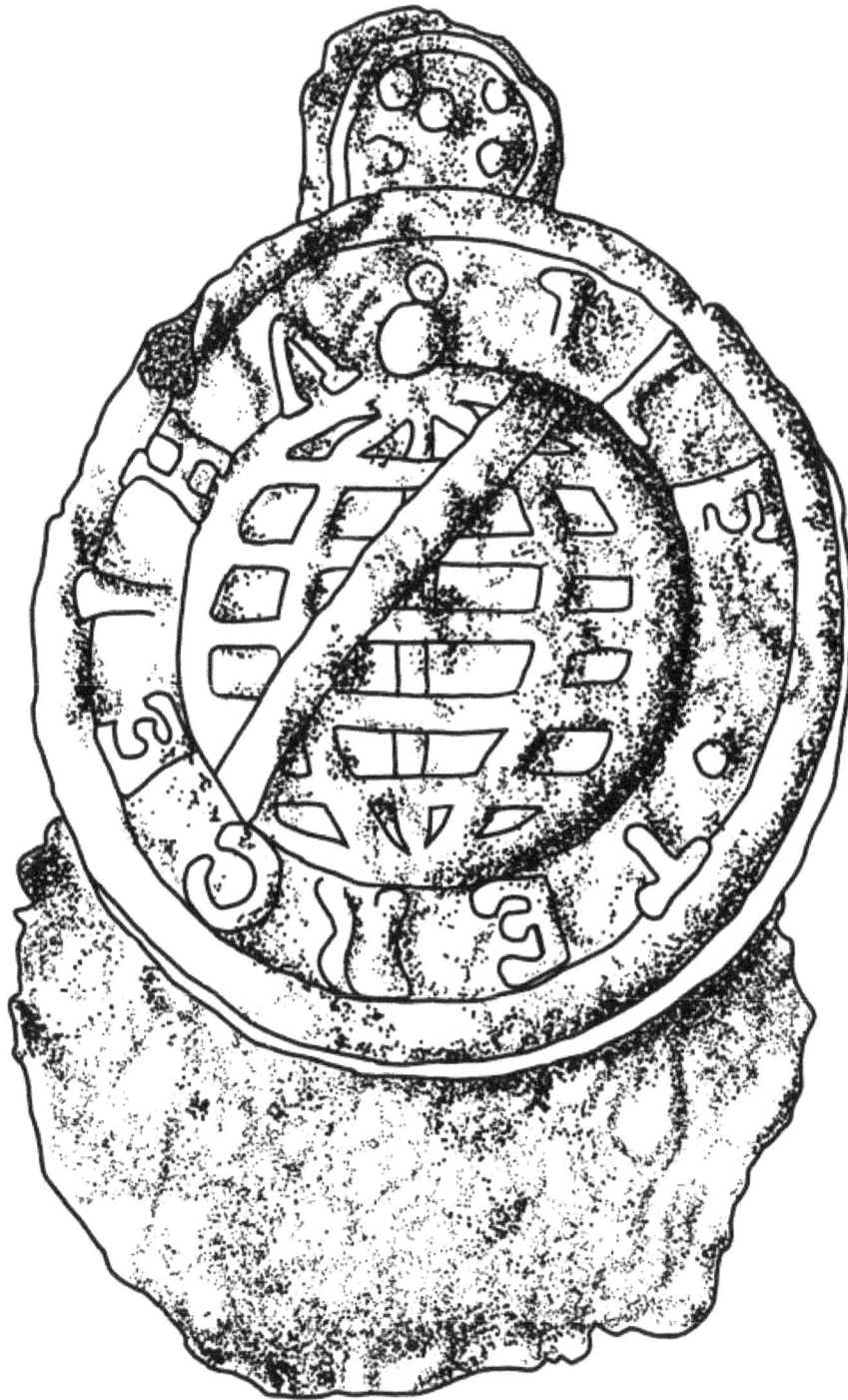

Category H: Artist's Perspective (Illustration)

First: Sketch of Terceira Lead Seal - Michelle Bouquet
Second: Illustration of Vessel 11: Thomas Hughes & Sons ironstone bowl by field school student Taylah Graham
People's Choice: Sketch of Terceira Lead Seal - Michelle Bouquet

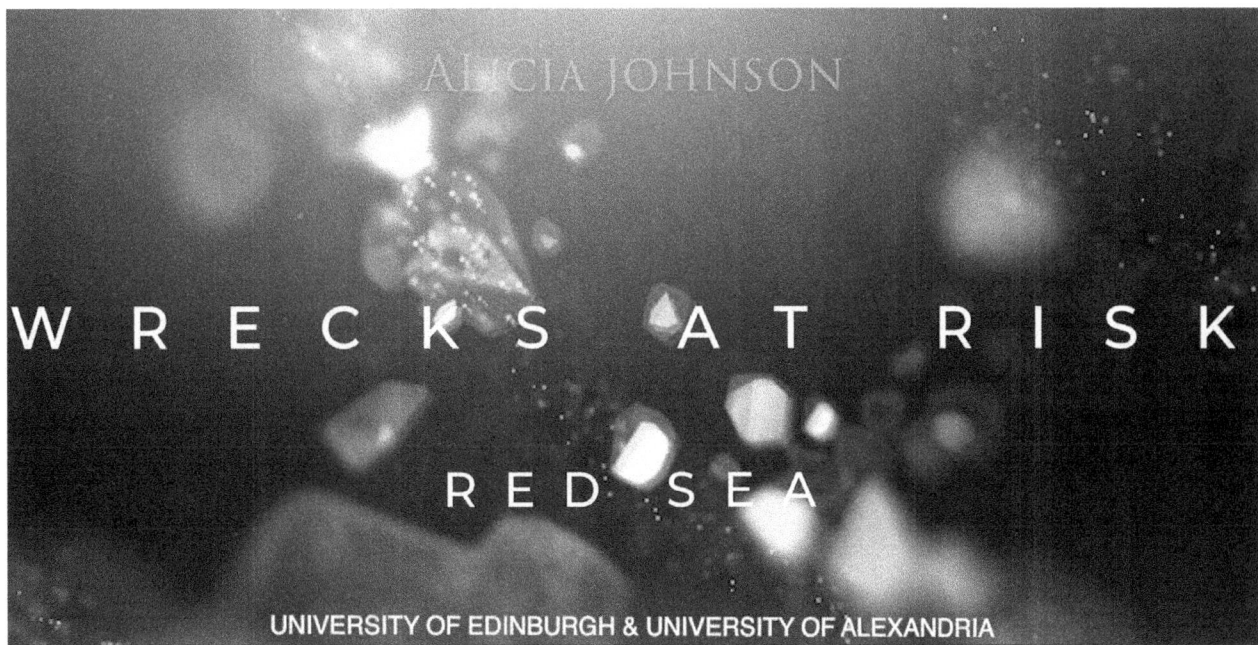

ALICIA JOHNSON

WRECKS AT RISK

RED SEA

UNIVERSITY OF EDINBURGH & UNIVERSITY OF ALEXANDRIA

Category I: Archaeological Video

First: Wrecks at Risk: The Carnatic – Alicia Johnson
Second: A Tale of Two Sites – Nicole Grinnan and Michael Thomin
Third: 16th Century Underwater Archaeological Investigation – Sarah Muckerheide
People's Choice: Wrecks at Risk: The Carnatic – Alicia Johnson